Synthesis Lectures on Engineering, Science, and Technology

The focus of this series is general topics, and applications about, and for, engineers and scientists on a wide array of applications, methods and advances. Most titles cover subjects such as professional development, education, and study skills, as well as basic introductory undergraduate material and other topics appropriate for a broader and less technical audience.

Fil A. Macias

Project Development Simplified

Design of Experience with a Focus on Abstraction

Springer

Fil A. Macias
University of Texas at El Paso
El Paso, TX, USA

ISSN 2690-0300 ISSN 2690-0327 (electronic)
Synthesis Lectures on Engineering, Science, and Technology
ISBN 978-3-032-01042-1 ISBN 978-3-032-01043-8 (eBook)
https://doi.org/10.1007/978-3-032-01043-8

This Springer imprint is published by the registered company Springer Nature Switzerland AG
The registered company address is: Gewerbestrasse 11, 6330 Cham, Switzerland

Experiences are the real essence of life, especially those shared with my wife, son, and daughter. These moments strengthen our bonds and become memories that last a lifetime.

Preface

Project development can be a complex process that requires careful planning and management. To simplify this process, systems engineering and project management techniques are often used. *Abstraction* is a crucial technique that allows project managers to focus on essential details while avoiding unnecessary complexities. Abstraction is used in various fields of science and in nature to simplify complex phenomena, create models, and gain deeper insights into something under investigation. By representing and understanding project development at various levels of detail and perspectives, project managers can identify essential features, properties, or behaviors that facilitate planning and managing project development.

Engineers often use *hierarchical abstraction* to create models or representations that capture the essential aspects of a system. Hierarchical abstraction involves breaking down complex systems or projects into manageable and organized levels or layers of abstraction. Each level represents a unique perspective or detail of the system, providing a better understanding, analysis, and management. This practice helps in systematically organizing and managing complex systems. This form of abstraction is a valuable technique in systems engineering and project management as it provides a structured way to understand, design, analyze, and manage complex systems or projects. The process promotes clarity, organization, and effective decision-making, leading to successful system development and project execution. These abstract representations further serve as a means of communication, enabling stakeholders to understand the system without being overwhelmed by its intricate details. By abstracting a system, engineers can create a more manageable and understandable representation, which in turn facilitates problem-solving, decision-making, and collaboration.

This technique is widely used in fields such as computer science, information architecture, and cognitive psychology. In computer science, it is used for designing complex systems in smaller modules or components. In information architecture, it helps to organize and structure content for easy navigation and understanding. In cognitive psychology, it is employed to understand how humans mentally organize and categorize information.

Abstraction is a fundamental aspect of understanding and interacting as described in the following examples:

1. Abstraction is a fundamental concept in nature that presents itself in various forms. For instance, in biology, the notion of a "species" is an abstraction that groups organisms with similar characteristics, enabling us to better understand and study them. Similarly, ecological models often simplify complex ecosystems into food chains or trophic levels to comprehend the flow of energy and nutrients.
2. Mathematics is a subject that heavily relies on abstraction. Mathematical concepts, such as numbers, sets, and functions, are abstractions that represent and generalize real-world phenomena. By abstracting complex ideas into simpler ones, mathematicians can create powerful models and theories with applications in various scientific disciplines.
3. In computer science, abstraction involves hiding the details of complex systems and creating simplified models that can be used to develop software and databases. Abstraction enables programmers to create higher-level programming languages that make it easier to write efficient and modular software, while database designers use it to create abstract data models that simplify complex data structures.
4. In the field of physics, abstraction plays an important role in creating mathematical models that describe the behavior of physical systems. These models simplify the complexities of real-world phenomena into mathematical equations and principles, enabling predictions and calculations. For instance, Newton's laws of motion provide a simplified abstraction of the behavior of objects in action, making it possible to predict their behavior and perform calculations.
5. In social sciences such as economics, sociology, and psychology, researchers often simplify complex phenomena into models to study human behavior, societal trends, and economic systems

Technical managers often rely on a rigid set of rules and procedures to solve problems, but it is important to encourage engineers to develop their own heuristics. These heuristics help them understand not only the process but also the technical elements that need to be incorporated into the solution. By developing heuristics based on abstraction, engineers can improve their decision-making process and save time and effort. Rather than analyzing every detail of a complex problem, readers can rely on simplified rules or guidelines derived from abstract knowledge. This enables them to make quick and reasonably accurate judgments, even in situations where time or resources are limited.

Furthermore, modern project development can be challenging. It involves the incorporation of complex systems with the technologies necessary to satisfy stakeholders' needs and requirements. This complexity associated with such an undertaking makes it challenging to understand and manage, leading to potential delays, errors, and cost overruns. Technologies are constantly evolving and changing. Keeping up with these changes and selecting the right technologies for a project can be challenging requiring continuous learning and staying updated. In addition, coordinating and collaborating effectively across different roles can be challenging. Modern projects involve teams of developers, designers, testers, and other stakeholders working together. That collaboration coupled with the need to handle large amounts of data, processes, dependencies, and constraints make for a challenge. Balancing this challenge with the need for high-quality deliverables can be difficult, as it may require prioritization, efficient resource allocation, and effective project management techniques.

Meeting the expectations of stakeholders therefore can be a formidable task, as it requires a deep understanding of their needs, iterative design and development processes, and continuous feedback and improvement. Modern project development is full of challenges, such as technical complexities, evolving technologies, collaboration needs, time and budget constraints, security concerns, and user expectations.

Successful project development necessitates various technical expertise, effective project management, and continuous learning and adaptation. Ensuring that the system can scale and perform well under heavy loads can also be challenging, as it requires careful architectural design, optimization, and testing.

To increase the chances of success and effectively communicate the process and intent to all stakeholders, the ability to navigate and interact with complex systems or concepts must be developed. This includes incorporating models or abstractions to communicate complexity and reduce ambiguity. Hierarchical abstraction is a good example of a tool that can be used to achieve this.

Organizing information in a structured and intuitive manner is crucial to its effective presentation. Hierarchical abstractions are a useful tool in achieving this goal. By categorizing and grouping information into different levels, users can easily navigate and access the content they need. For instance, in a file management system, hierarchical folders provide an efficient way to organize and locate files based on their hierarchy.

Progressive Disclosure: Hierarchical abstractions allow for progressive disclosure of information, revealing details at different levels of abstraction as needed. This approach prevents overwhelming users with too much information upfront and helps them focus on the most relevant content. For example, in an e-commerce website, product listings can initially show high-level information, but users can delve deeper into the product details by navigating through hierarchical layers.

Navigation and Exploration: Hierarchical abstractions can guide users through complex systems by providing a straightforward navigation structure. It helps users understand their current position within the system and provides a roadmap to explore further. Interactive menus, breadcrumbs, or tree structures are examples of hierarchical navigation elements that aid users in exploring and finding desired information.

Contextual Awareness: Hierarchical abstractions can provide contextual cues and visual indicators to help users understand the relationships between elements or concepts. Visual representations, such as nested folders or diagrams, can convey hierarchical structure and highlight the relationships between components. This enhances user comprehension and aids in decision-making.

Customization and Personalization: Hierarchical abstractions can facilitate customization and personalization by allowing users to modify or rearrange elements at different levels. Users can prioritize or hide certain information based on their preferences, tailoring the experience to their needs. Customizable dashboards or interface layouts exemplify how hierarchical abstractions can enable personalization.

Task and Workflow Management: Hierarchical abstractions can support task and workflow management by breaking down complex processes into smaller, manageable steps. Users can navigate through different levels to access relevant actions or tools required to complete a particular task. This simplifies complex workflows and enhances user productivity.

Error handling and troubleshooting can be challenging in complex systems. However, hierarchical abstractions can simplify the process by isolating issues to specific levels or components. This allows for more targeted error messages or alerts, providing users with actionable information to resolve the problem. By clearly identifying the source of an issue, it becomes easier to diagnose and solve it efficiently. Therefore, the use of hierarchical abstractions can greatly improve error handling and troubleshooting in complex systems.

This narrative intends to focus on systems engineering from the management perspective. Systems engineering is not just a subject; it's a mindset. It's not about following a perfect script. It's about learning, adapting, and evolving. There is no step-by-step recipe for success like following a process in a cookbook. It is in understanding what matters and having the strength to let go of what doesn't. Embracing challenges rather than avoiding them is essential, as that is where growth happens. That is where innovation is realized, or rather, where the experience is designed to make it possible.

In general, designing experiences through hierarchical abstractions aims to offer users structured, organized, and intuitive interfaces that simplify complexity, facilitate navigation, allow customization, and enhance comprehension and productivity. By utilizing the benefits of hierarchical structures, this approach enables users to interact with complex systems or concepts efficiently and flexibly.

El Paso, USA Fil A. Macias

Competing Interests The author has no competing interests to declare that are relevant to the content of this manuscript.

Contents

1 Design of Experience

The quote by Keats suggests that proper understanding and meaning come only through personal experience. In other words, knowledge gained from direct experience is more profound and authentic than anything learned indirectly. Instead, ideas, observations, generalizations, or abstractions can be relied upon, allowing information to mature, adapt, and support decision-making. Managing a project or developing a system involves making decisions.

The intent is to contrast between different modes of acquiring knowledge, such as through direct personal experience, through learning in a classroom, or perhaps by observation or through the derivation of abstract concepts that can be applied to a range of different situations [1].

When something is experienced, the focus is to engage with its nuances, complexities, and perhaps emotional aspects, creating a level of understanding that cannot be fully captured through second-hand means like reading about it or by hearing from others [2].

Suppose the challenge is to develop a new project about a subject that is novel and not fully understood. The focus then becomes on how to adapt to the complexity and ambiguity associated with such a development. Perhaps borrowing from experiences gathered from previous projects could help lessen the ambiguity and improve our understanding based on the following [3]:

- Development of Ideas: Conceptual frameworks that help us interpret our new reality.
- Observations: Learning from experiences and actions, not necessarily all our own.
- Generalizations and abstractions: Describe processes of facilitating a specific problem to help augment our understanding by focusing on essential factors of a specific problem.

F. A. Macias, *Project Development Simplified*, Synthesis Lectures on Engineering, Science, and Technology, https://doi.org/10.1007/978-3-032-01043-8_1

Not everything is known at the start of a new project, especially one aiming to create something new. This uncertainty drives innovation and challenges decision-making, planning, and management. To navigate unknowns, Project managers and system developers must find ways to make informed choices, adapt to new situations, and broaden their perspectives. Ambiguity can hinder understanding, making it necessary to establish methods for expanding knowledge. Since failure is not an option, a structured approach to learning and decision-making must be included as part of the development process.

1.1 The Adjacent Possible

In the article by Steven Johnson titled The Genius of the Tinkerer,

The discussion focuses on how innovation will manifest itself from the collective network of ideas that may have been inherited or simply stumbled across as they are put together into some new form. This initial stumbling, which is characteristic of new project developments, was referred to as "the adjacent possible." A phrase that is useful since it serves to capture both the limits and the creative drivers necessary for change and innovation. In the case of systems engineering, the adjacent possible serves to describe potential alternatives to define an achievable solution; it is a term that refers to looking at all the ways in which the problem can evolve towards a solution, opening up the possibility of a new potential [4].

With each new project, challenges and assumptions will cloud viewpoints and test our abilities to understand the system complexities. Self-limiting perspectives on ideas should be avoided since this makes it difficult to explore possible alternatives. The intent should be to find ways to improve understanding of the associated complexity through a variety of means, such as abstraction/modeling, experimentation, prototyping, and abstraction.

Modern system development requires understanding complexity, as systems today are highly interconnected, dynamic, and nonlinear. The connectivity between components may be needed for operation, but may result in unpredictability.

Traditional linear approaches of development often do not capture inherited attributes of system complexity, multidimensionality, and dynamic characteristics. A more effective approach would be to find ways to understand system nonlinearity, feedback loops, emergent behaviors, and network dynamics [5].

Systems may exhibit emergent properties, which are unplanned or unexpected features that result from interconnected components and/or processes. These must be managed from a whole system perspective rather than focusing solely on individual components because the whole delivers the desired capability. The challenge lies in comprehending the entire functionality based on the intricate and dynamic nature of the individual elements. Critical features of complexity may include [6]:

- Interconnectedness: Components depend on and influence each other.
- Unpredictability: System behaviors may not always be known or easy to understand.
- Emergence: Unplanned system behaviors (functions) may develop and cannot be readily predicted by examining individual parts.

The relationship between components presents the hierarchical nature of the system. Which can be described as a structured framework to define and analyze system functionality and determine how the system will be developed and integrated into its operational context. To clarify, the operational context refers to the specific conditions within which a system operates. Understanding external and internal factors reveals how they influence system performance, either enabling or constraining it. Rather than fragmenting the system, the focus should be on identifying key leverage points, modeling interactions, and uncovering hidden outcomes [7].

A structured model should leverage principles of systems thinking to enhance clarity, adaptability, and resilience. By mapping out system components, interactions, and feedback loops, such a model helps identify leverage points, which are planned areas where directed interventions yield significant improvements.

These leverage points might include regulatory mechanisms, resource allocation processes, or key decision-making nodes, depending on the system being analyzed. Structured models also help manage complexity and ambiguity by making implicit relationships explicit, facilitating scenario analysis and forecasting [8].

The intent is to find clarity in addressing complex situations to help system developers better understand how systems evolve and adapt to changing needs and requirements. Through this approach, interactions, dependencies, and feedback loops become more transparent, facilitating informed decision-making, mitigating risks, and enhancing adaptability. Ultimately, this framework should support resilient, efficient, and well-integrated systems that align with long-term objectives and stakeholder expectations and mitigate different types of unintended outcomes in complex systems [9].

A systems engineering approach is necessary to manage risks, using predictive modeling, risk analysis, and adaptable design to foresee and prevent potential issues. System developers must carefully examine how different system components interact, identify weak points, and implement flexible controls to maintain reliability and efficiency. Engineers can enhance predictive capabilities, optimize decision-making, and maintain consistency in addressing complexity by leveraging computational modeling, adaptive frameworks, and integrated system architectures. For example, consider the following different types of unintended outcomes in complex systems:

- Feedback Loops: These are fundamental concepts in systems thinking that illustrate how a system's output can inadvertently influence its input, either by reinforcing or counteracting its behavior. While these loops are essential for maintaining stability, they can also lead to unintended consequences when poorly designed or implemented [10].

Consider induced demand as a common trend in transportation planning. The solution often considered is to find ways to alleviate traffic congestion by building more roads; however, this usually leads to increased demand. This happens because motorists are encouraged to drive more, which results in an increase in traffic congestion, resulting in a self-defeating cycle where the solution worsens the original problem. This circumstance's outcome emphasizes the need for a more comprehensive approach to transportation planning that considers factors beyond expanding infrastructure. The issue is that congestion levels may return to their previous state even after road widening, negating any benefits gained. For example, consider the following [11]:

- Emergent Behaviors: A defining characteristic of complex systems is the interaction among individual components, which can result in emergent behaviors that are not predictable based on just looking at the functionality of the components alone. For example, social media algorithms, designed primarily to enhance user engagement, may unintentionally create echo chambers, amplify misinformation, or reinforce biases and outcomes that were not explicitly programmed but emerge from the dynamic interplay between user behavior, content distribution, and feedback loops within the system. This highlights the importance of systems thinking in anticipating and addressing unintended consequences in complex, adaptive environments [12, 13].

A study by DuckDuckGo found that if online platforms adjusted their algorithms to make content less visible or prominent, it reduced the potential of misinformation by more than 50%, highlighting the significant influence of algorithmic design on the flow of online information platforms. This finding underscores the need for platforms to shift from informational interventions to algorithmic adjustment, since the modification of content mechanisms can have a more direct and effective impact on what information is made available. By embedding systemic controls with appropriate recommendations, platforms can proactively mitigate the unintended effects of engagement-driven algorithms and promote a more reliable and balanced digital information space [14, 15].

Development is challenging, and enormous amounts of data must be gathered, synthesized, and understood to help plan and develop projects. As Danelle Meadows explains in Thinking in Systems, an effective manager can clearly define the problem, identify the structural components driving it, and pinpoint leverage points: strategic areas where small adjustments can trigger significant systemic change [16].

Mitigating those triggers for change necessitates deeply comprehending the system's complex structure, dynamic behaviors, feedback loops, and inherent nonlinearities [17]. It involves understanding the bigger picture and addressing root causes instead of alleviating symptoms [18]. Developing a structured model or framework is, therefore, key to understanding and optimizing system behavior and ensuring adaptability in dynamic environments to support the following [19]:

- A Framework for Analysis: A well-defined model provides a systematic approach to analyze the system and understand its components and interactions.
- Knowing how to Optimize System Behavior: By understanding the system's inner workings, the model enables optimization for performance, reliability, risk, and other important metrics.
- Enhancement of Adaptability and Resilience: The model assists in anticipating and responding to changes in the operational context, enhancing the system's adaptability and resilience.
- Managing Complexity: A model aids in managing complexity by offering a means to visualize the system, comprehend its relationships, and analyze its behavior. It also facilitates decision-making regarding what is significant and what is not, which can help lessen the system's overall complexity. For example, traditional risk management strategies, which are typically reactive and based on pre-defined assessments, are insufficient to manage the complexity and unpredictability of contemporary projects.

Using "design of experience" as a metaphor for project development implies that framing the process of planning, developing a project, and desired outcomes is based on the developers' experiences and the impact on stakeholders, benefiting from the realization of a needed solution. The experience lends to how the project is done, but then changes the perspective of the Stakeholder, hopefully in the best possible way. The experience carries information and knowledge, providing insights to help plan and advance a project, and the other experience centers of the Stakeholder benefit from the solution, thus positively changing their outcome [20].

The experience is then two-dimensional, involving everyone, including the development team, end-users, and stakeholders. To manage this, it's essential to identify key touchpoints and interactions that shape this experience.

To truly focus on the user, it is essential to step into their shoes and those of all stakeholders. Understanding their needs, expectations, and potential pain points throughout the project lifecycle is crucial. This demands a deep analysis of both the solution and the problem space, driving genuine comprehension and empathy.

The "designing an experience" metaphor emphasizes clarity, engagement, and efficiency. Things don't happen by magic; practical design demands streamlined workflows, clear communication, and strong collaboration. Notably, user experience is not static; it is a dynamic process that evolves over time. This perspective naturally promotes an iterative development approach, where feedback is continuously gathered and integrated to enhance the overall experience. By embracing this mindset, teams can align more effectively on goals, priorities, and processes, resulting in stronger communication, smoother collaboration, and more informed decision-making.

In the paper: Experience Designs' Transformation Towards Experience-Driven Transformation: A Practical Perspective by Michael Lai, the analysis outlines the evolution of UX design into a broader experience strategy. This transformation is defined through four distinct phases [21]:

- UX to Experience Design: Transitioning from designing digital experiences to creating holistic, multi-touchpoint experiences.
- Experience Design to Experience Strategy: Shifting focus towards crafting the dynamic connection between users and the brand.
- Experience Strategy to Experience Management: Moving towards the quantification and data-driven analysis of experiences.
- Integration into Experience-Driven Transformation: Evolving into strategic consulting to foster experience-driven transformation.

Each phase is characterized by specific models and frameworks designed to address unique challenges:

- Domains of Experience: Foundational understanding of experience touchpoint, which refers to any point of interaction between a user and a system, such as a process step, or a support service.
- Holistic Experience Model: Comprehensive mapping of user interactions.
- Brand Experience Index: Quantitative assessment of brand interactions.
- Experience-Driven Transformation Model: Strategic guidance for experience-led change.

These models facilitate the shift from design to strategic experience management, driving meaningful transformation. The progression through each phase adds value and builds the foundation for the next stage, emphasizing that impactful design must be strategically grounded to enable transformation.

Heuristics, shaped by experience, acquired knowledge, and perhaps second nature, can enhance project planning and development. By addressing both the problem space and solution space, based on existing knowledge and identified needs, project development can better align with stakeholder expectations. This approach streamlines processes, enhances communication, and accelerates project delivery. Viewing the project as an "experience" fosters collaboration and innovation, helping teams identify opportunities for improvement [22].

Systems engineering in complex projects requires understanding the dynamic relationship between design and requirements. Rather than following a linear process where requirements dictate design, it's an iterative interchange. Design choices at higher abstraction levels impose requirements on lower-level components, creating a hierarchical structure where one set of design decisions becomes another's requirement [23].

Suppose the requirement is to build a car transmission for our new Super XYZ vehicle, for example. The overall automobile design determines the vehicle's weight, which in turn dictates the transmission's required output. In this way, a high-level design choice becomes a fundamental requirement for the transmission subsystem. This creates a hierarchy where one set of design decisions becomes the starting point or requirement for the next [24].

This hierarchical nature of system development is often more "middle-out" than strictly top-down. Compatibility with existing architectures and the availability of specific components can significantly influence high-level requirements. This challenges the traditional sequential approach to systems engineering, requiring a more integrated model where requirements account for hierarchical system construction, and architectural design notations better support component specifications. Failing to consider these interdependencies can lead to significant integration challenges later in the development process [23].

To mitigate these challenges, formalizing requirements at various levels of architectural abstraction is necessary. This approach is enabled by the virtual integration of components, allowing systems engineers to validate system-level requirements based on those allocated to individual components. Formalization enhances communication, streamlines processes, ensures alignment with project goals, and helps mitigate risks. By structuring information to reduce ambiguity and clarify dependencies, developers can enhance consistency and facilitate seamless integration throughout the development lifecycle [23].

Tracing requirements across different levels of abstraction is essential for identifying integration errors, even when individual components meet their functional specifications. Issues often arise during integration, making requirement traceability critical for detecting and resolving potential problems early [25].

Experience helps frame the varying levels of abstraction required to understand business, user, and system requirements, which specify detailed capabilities, behavior, and performance within their respective development contexts.

This identifies the environment, constraints, and conditions in which a system, product, or project is designed, built, and implemented. Business requirements outline the overarching objectives and expected value of the project, while user requirements translate these goals into specific, actionable needs from the end-users' and stakeholders' perspectives [26].

Furthermore, the interdependence of requirements and architectures must be actively managed. In the early stages of co-design, requirements and architecture can be relatively informal and fluid, resulting in a flexibility that fosters innovation and risks inconsistencies and integration errors [23]. Establishing transparent processes for managing change, documenting decisions, and communicating updates to all stakeholders is crucial for ensuring that requirements and architectures remain aligned throughout the project lifecycle [27].

The goal is to establish a development environment where innovative requirements and architectural solutions evolve together, ensuring seamless integration and fostering innovation [26]. Managing complexity in modern systems engineering requires a systematic approach that acknowledges the hierarchical nature of systems and formalizes requirements at various levels of abstraction. Experience, abstraction, and hierarchy are not merely buzzwords; they are valuable tools that assist in developing complex systems. Like an expert chess player who anticipates multiple moves ahead, an experienced systems engineer should utilize these principles to navigate the complexities of modern projects, ensuring that all system elements function as intended toward the desired outcome [23].

An understanding of hierarchical system structures is often gained through experience [28]. An engineer doesn't just learn these concepts from a textbook, they are developed through real-world projects, shaped by both successes and failures. This hands-on experience enables engineers to break down a system into manageable, meaningful components, much like analyzing a complex machine by examining each moving part [29].

Consider the development of a modern aircraft. An experienced systems engineer recognizes that an aircraft is not a single, monolithic entity but a hierarchy of interconnected systems, such as power, actuation, and lubrication systems, each with its requirements and constraints.

Modern aircraft development is a complex process that requires precise planning, seamless integration, and rigorous testing of numerous subsystems. An aircraft is not a singular entity but a complex network of interdependent subsystems such as power distribution, actuation, lubrication, and more, each with its complexities. The focus is on understanding the aircraft's structure, dependencies, interdependencies, and hierarchical constructs to assess a hierarchy of requirements and how they need to evolve [30].

A modern aircraft is systematically divided into primary systems, which are further segmented into subsystems and components. This can be characterized as a hierarchical structure that facilitates design clarity and supports modular development. The significant subsystem categories may typically include [31]:

- Power Systems encompasses engine performance, power distribution, and energy management. Efficient power systems are critical for both propulsion and onboard electronic functions.
- Flight Control Systems include actuation mechanisms, control surfaces, and feedback loops responsible for maintaining aircraft stability and maneuverability. Modern flight control systems are often fly-by-wire, where electronic signals replace mechanical linkages.
- Fuel Systems: Managing fuel storage, distribution, and monitoring engine maintenance is critical to maintaining engine performance.
- Lubrication Systems: Aircraft consist of many moving parts, requiring lubricants to reduce friction and wear. Lubrication systems are necessary to ensure the longevity and reliability of the engine and moving components.
- Avionics Systems: These systems address communication, navigation, and sensor integration, allowing pilots to operate the aircraft safely and efficiently.

- Environmental Control Systems (ECS): Manage cabin pressurization, temperature regulation, and air quality to maintain passenger comfort and safety.
- Safety and Redundancy Systems: Modern aircraft are equipped with multiple layers of safety protocols, fail-safes, and redundancy checks to prevent single points of failure.

Although each system has its own specific set of requirements, they are deeply interdependent. For instance [31]:

- The actuation system for control surfaces relies on power availability from electrical or hydraulic systems to function effectively.
- The lubrication system is tied to the engine's thermal management, affecting overall efficiency and preventing overheating.
- Power systems also supply critical energy to avionics and navigation systems, ensuring the reliability of real-time data processing and communication.

Such interdependencies mean that engineers must account for cascading effects; a fault in one system could propagate across others if not properly managed. Systems engineering is the backbone of modern aircraft development. It adopts a top-down, bottom-up approach to design [32]:

- Top-down: High-level requirements are defined first, which then drive the design of subsystems. Engineers outline performance criteria, safety standards, and integration benchmarks at the conceptual phase.
- Bottom-up: Individual components are tested and validated before integration. This ensures that when subsystems are connected, they operate harmoniously.

They understand that design choices at higher levels of the hierarchy, such as selecting a specific actuator, impose requirements on lower-level system components [23]. This understanding enables engineers to proactively manage dependencies, anticipate challenges, and ensure the system operates efficiently. Without experience-driven insights, such a project would likely linger and fail [33]. Abstraction is key for systems engineers to manage complexity. It's about simplifying complex systems into higher-level concepts while concealing non-essential details [33].

This is not about simplifying things, but rather focusing on essential concepts. A system can be structured hierarchically, consisting of subsystems, components, subcomponents, and parts [32]. By using abstraction and hierarchy, models for SE can be scaled and simplified by decomposing the system into manageable and meaningful components, or rather, those that enable a clear, essential purpose in the system's structure and function [32].

Abstractions focus on business and organizational models at the enterprise level, while functional architectures are emphasized at the systems level. To prevent overly generalized models, irrelevant features should be avoided, and essential elements should be adequately distinguished from nonessential ones [34].

This approach allows engineers to focus on specific functionalities without being distracted by unnecessary details. Consider driving a car: the driver doesn't need to understand the intricacies of the engine's mechanics but knows how to interact with the steering wheel, pedals, and controls. It involves identifying the right level of detail for the given objective, a principle every experienced engineer understands [35].

The capability to formalize requirements across various levels of abstraction is a key characteristic of an experienced systems engineer [36]. This process involves translating high-level stakeholder needs into definite, verifiable specifications for system elements. Formalization is not a one-time event but an iterative refinement, shaped by experience and a growing understanding of the system. As system verification and validation evolve through iterative testing, our understanding of best practices improves through continuous analysis and feedback [37].

Different levels of abstraction allow for organizing various models, such as business and organizational models, functional architectures, and software designs [34]. Understanding the interaction between procedural and declarative knowledge is fundamental to formalization. This supports a more effective and comprehensive management strategy. Consider the following definitions [38]:

- Procedural Knowledge: This refers to knowing how to do something. Addresses the ability to perform tasks and processes, such as riding a bike, solving a math equation, or programming a software function. This kind of knowledge is often implicit and gained through practice.
- Declarative Knowledge: This is about knowing what to do. This requires having facts, information, and concepts in mind, such as knowing the capital of Spain, the definition of a term, or the rules of chess. This type of knowledge is typically straightforward to convey.

While abstraction offers significant benefits, it also presents inherent challenges. Over-abstraction can lead to pointless generalization, where essential details are lost, making implementation difficult. Conversely, under-abstraction can result in excessive complexity, making the system cumbersome and hard to manage [34].

Furthermore, poorly chosen abstractions can lead to incorrect analysis [39]. Experience teaches systems engineers the importance of balancing abstraction, finding the middle ground between unnecessary generalization and overwhelming complexity [34].

Experience, abstraction, and hierarchy are deeply interconnected in systems engineering. Experience builds the foundation for understanding complex systems, abstraction provides the tools to manage that complexity, and hierarchy structures and relates system components. Mastering these principles enables systems engineers to tackle modern project challenges and deliver innovative, adaptable solutions that meet evolving needs.

Technology will continue to advance; however, the core principles of systems engineering will remain to guide those who strive to create not just a "better mousetrap" but more reliable, efficient, and scalable systems [36].

1.2 Abstraction and Hierarchical Understanding in Complex Systems

Organizing information into a functional structure remains important for systems engineering and project management. It is not as simple as copying previous methods, but rather a way of deriving information to aid in decision-making and guide the system's or project's development [40]. Effective project organization prioritizes resources and reduces disruptions [41].

Effective systems engineering and project management rely on well-structured information to minimize ambiguity. System developers enhance their understanding of project requirements by organizing information systematically and defining relational dependencies. This structured approach ensures that objectives are clear, key details are addressed, and the system is efficient and well-integrated.

By focusing on essential goals, developers can establish clear, realistic requirements that align with the projects' aims, guiding the development process effectively [42].

Abstraction simplifies complexity by minimizing nonessential details and providing a higher-level perspective. This enables developers to concentrate on specific functionalities without being sidetracked by extraneous information [43].

In computer programming, abstraction manages a system's underlying complexity while still providing the necessary functionality to the user. A common misconception is that abstraction inherently obscures complex processes or hides too much information. Instead, abstraction encapsulates complexity within a defined boundary, allowing users to interact with the system at a higher level without needing to grasp its internal intricacies. Abstractions can be described in terms of their level, scope, and objectives to evaluate the maturity of engineering processes without focusing on specific details [44].

Recognizing that not all answers are immediately available is crucial when working on complex projects. In computer programming, for example, abstraction manages underlying complexity while ensuring the user can access necessary functionality. The goal is to enable programmers to focus on higher-level operations without being concerned with the intricate mechanics running in the background. At its core, abstraction reduces complexity by presenting a streamlined view of a problem, making it easier to manage, understand, and solve. Practical abstraction should focus on the essential goal without becoming entangled in unnecessary details [45].

Abstraction offers significant advantages, but it becomes over-engineered and brittle when it is too vague. The result is a hierarchical relationship between critical system elements, such as functionality, interfaces, and complexity, that is not readily apparent [46].

Hierarchical relationships, which structure systems into layers and simplify complex interactions, reflect the need for structured approaches to problem-solving. Dave Snowden's Cynefin Framework provides a model for categorizing problems into five domains: simple, complicated, complex, chaotic, and disorder (Confused/Aporetic) [47]. These are defined as follows:

- Simple (Obvious): Cause-and-effect relationships are clear, and best practices can be directly applied.
- Complicated: Cause-and-effect relationships exist but require expertise and analysis to understand.
- Complex: Cause-and-effect relationships are only identifiable in hindsight, necessitating experimentation and adaptive learning.
- Chaotic: Cause-and-effect relationships are indiscernible, requiring immediate action to restore order.
- Disorder (Confused/Aporetic): The situation is unclear, necessitating further information to determine the appropriate domain.

This framework helps decision-makers assess uncertainty and select the most effective strategies for managing various types of problems. By employing this framework, systems engineers can more effectively navigate the complexities of hierarchical structures and make informed decisions based on the nature of the issue [47]. These relationships are fundamental for developing a systematic approach to understanding how complex systems interact while maintaining simplicity and clarity. System developers apply various strategies, such as top-down or bottom-up methods, in a coordinated manner to effectively manage complexity. Project managers can enhance efficiency throughout the project lifecycle by emphasizing the interactions between system components [47].

In the complicated domain of the Cynefin Framework, hierarchical relationships help experts systematically analyze and resolve problems where cause-and-effect relationships are known but require specialized expertise. System engineers are well-equipped to navigate this domain, coordinating the necessary actions to ensure project success. This structured approach facilitates the use of quantitative techniques and value stream mapping, aiding in the management of systems with multiple layers of impact [48].

Abstraction simplifies complexity by emphasizing higher-level perspectives, a key method for navigating the Cynefin Framework. Engineers benefit from adopting a high-level perspective, allowing them to make meaningful contributions while maintaining clarity in complex projects [48].

Within the simple domain, abstraction categorizes problems with straightforward solutions, enabling responses based on established best practices. Developers may reference previous projects and known resolutions to devise an efficient new course of action, thereby streamlining decision-making by focusing on the problem's characteristics. Conversely, in the chaotic domain, abstraction imposes constraints that stabilize the system, allowing for deeper analysis to occur. This structured approach facilitates decisive action in crises, providing order amid uncertainty [48].

The Cynefin Framework helps system engineers understand complex problems by categorizing them into distinct domains based on their characteristics [49]. By acknowledging the systems' dynamic nature, engineers understand that solutions must evolve alongside changing contexts. This awareness enables them to design systems with built-in flexibility and adaptability, ensuring they can effectively respond to uncertainty and unforeseen changes [47].

Within the Disorder domain, the process involves abstraction and hierarchical decomposition to break down unclear problems into manageable parts [48]. In managing the disorder domain, the complexity of the problem must be broken down into manageable parts to prevent systems engineers from becoming overwhelmed. As each aspect is deconstructed into smaller components, abstraction helps by identifying their relationship to the broader objective [47].

By combining hierarchical relationships and abstraction, the Cynefin Framework provides a structured yet adaptable approach to managing complexity across various domains. This synergy enhances systems engineering and organizational decision-making by helping identify critical system characteristics. One of the framework's key strengths is enabling system engineers to adapt to diverse project challenges while guiding leaders in making informed decisions, even in complex environments [47].

Overall, the Cynefin Framework strengthens the ability to navigate complexity, make informed decisions, and foster adaptability, leading to improved system outcomes. A structured decision-making framework helps project teams execute initiatives more effectively. Since systems have many interconnected components, each project element can have a direct impact on outcomes. A deeper understanding of these interdependencies enables teams to make more strategic and well-informed decisions [47].

Effective system design depends on managing complex interconnections, which requires a clear understanding of how each part relates to the whole. Hierarchical models, which organize systems into structured levels based on significance, are crucial in addressing complexity. By layering or modularizing systems, these models clarify relationships and dependencies, making interactions between components more transparent and understandable [50].

1.3 Hierarchical Structure: Organizing System Components

Hierarchical structures arrange system elements into ranked levels, revealing the connections between higher-level functions and lower-level components. This structure facilitates a better understanding of the system and improved design efficiency. When tackling complex projects, simplifying and breaking down key elements prevents them from becoming overwhelming. A well-defined hierarchy facilitates decision-making by providing a structured approach to evaluating options, particularly in uncertain conditions.

Hierarchical organization enhances modularity by breaking down systems into smaller, well-defined components, making the system's organization and functionality more easily understood. A top-down approach helps manage development by providing insights into potential issues that may arise during implementation [50].

Abstraction simplifies complexity by hiding unnecessary details and presenting a higher-level perspective. This allows developers to focus on specific functionalities without being distracted by extraneous information. Abstraction is a cognitive process that simplifies complex ideas by focusing on their essential aspects. This involves generalizing information to create an overarching understanding. The inner details and complexity of the underlying components are concealed by abstracting the system. Creating layers of abstraction establishes clear boundaries between distinct parts of the system, making it easier to modify components without needing to understand intricate details. As engineers use abstraction in conceptualization, they can better implement and assess system design [50].

Effective development management is essential for system development, ensuring that components remain loosely coupled and enabling scalable, adaptable systems where changes can be made without compromising overall functionality or integrity. Beyond simply identifying connections, tracing ripple effects across components is crucial for understanding system-wide impacts.

Managing these dependencies involves assessing risks and vulnerabilities while developing contingency plans to mitigate potential disruptions. Additionally, teams must acknowledge that a required solution may not always exist, necessitating the allocation of time and resources for its development. Proactive planning and diagnostics provide valuable insights into potential challenges, limitations, and risks that may arise throughout development [50].

1.3.1 Balancing Abstraction and Hierarchy

The practical application of abstraction and hierarchical analysis requires balance. Modular design facilitates this by structuring systems into layers or modules, allowing modifications without disrupting the entire system. A layered architecture separates concerns by defining clear interfaces between system components, ensuring that high-level objectives and critical details are correctly addressed, leading to a well-integrated and efficient system.

Additionally, the ability to replicate components supports seamless scalability. System engineers can enhance long-term planning by continuously monitoring and analyzing system health and performance, anticipating equipment lifecycles, and estimating maintenance requirements to ensure reliability and sustainability [50].

Combining abstraction with hierarchical models offers a comprehensive approach to systems engineering. This enables engineers to navigate complexity, refine system design, and make informed decisions. By focusing on essential elements, ensuring clear communication, and maintaining adaptability, organizations can achieve more effective project execution [50].

Recognizing system complexity is essential for developing solutions that adapt to changing contexts and requirements. Abstraction, when paired with a hierarchical model, offers a structured approach by organizing system components into ordered levels based on importance, functionality, or shared characteristics. This layered structure reduces complexity by dividing systems into manageable modules, clarifying relationships, and defining dependencies. Hierarchical design also enhances scalability, enabling the smooth integration of new features or modifications without disrupting the entire architecture.

1.3.2 Defining Dependencies in Project Management

Dependencies in project management involve managing and scheduling tasks while adhering to their sequences and requirements. According to the Project Management Body of Knowledge (PMBOK), every task should have a dependency because, in project management, all tasks are interrelated. The Project Interdependency Management practice (PIM) integrates processes, tools, and techniques that allow project managers to identify, validate, analyze, track, and report on external project linkages [51].

Recognizing and managing dependencies means ensuring that the key connections between system components are clear and well-maintained. Every project presents its own set of challenges, which are often not immediately apparent. Over time, experience plays a crucial role in enabling engineers to make informed decisions, ultimately leading to more robust systems and fewer problems in the long run [52].

1.3.3 Key Principles of Dependency Management

Dependency management involves identifying, anticipating, and controlling dependencies among tasks, people, processes, and systems [53]. Effective dependency management does not enforce rigid structures but fosters an understanding of system relationships. Experience is vital in project planning, as it enables developers to predict and address challenges effectively. Continuously improving management and development capabilities is essential for successfully navigating complexities and achieving efficient outcomes [52].

Effective project development relies on well-crafted recommendations that support decision-making and solve problems. Good management is all about making informed decisions, and that starts with designing an experience that brings clarity, usability, and efficiency to the process. When stakeholders and developers interact smoothly with the system, it strengthens decision-making, streamlines problem-solving, and keeps project outcomes aligned with both technical and business goals.

Prioritizing user experience enables project developers to build intuitive systems that foster collaboration, enhance engagement, and facilitate successful execution.

1.4 Models and Abstractions

Models are representations of real-world objects or systems, while abstractions simplify complex information by focusing on fundamental details and suppressing unnecessary ones [54].

Several established system development models, like the Waterfall, Spiral, and Vee models, offer structured approaches. Each model has its strengths and weaknesses, and the choice depends on the specific project and its requirements. Agile engineering methods provide a more iterative approach, emphasizing continuous feedback and adaptation to change [55].

The Systems Engineering Vee Model is considered a framework that shows the connection between system development, validation, and eventual deployment into its operational setting. The left side focuses on breaking down and designing the system, while the right side ensures it meets requirements through testing and integration. The Vee helps simplify and visualize the complex, detailed process of systems engineering into a clear, structured diagram. By illustrating how design and testing activities interrelate across a project lifecycle, project developers can more effectively manage complexity without getting bogged down in every technical detail. Essentially, it is a conceptual model of the process, which is an abstraction of the actual process.

Abstraction is one way to manage complexity in the models [56]. By emphasizing a system's core characteristics while filtering out irrelevant details, abstraction enables developers to work at different levels of detail without becoming overwhelmed by overall complexity. This approach simplifies complex systems, making them easier to analyze, manage, and understand [57].

The Vee model is a central framework in systems engineering that represents the development of a system from defining the problem to validating the solution. As both a hierarchical breakdown and a structured abstraction, the Vee model can be used to represent the systems engineering process by showing how system development is decomposed into layers of detail and then progressively integrated and verified to form the whole system. This version highlights iteration, showing how each step on the left, from understanding needs to designing the system, has a matching step on the right to verify or validate what was built.

The Problem Space focuses on what the system should do. The Solution Space focuses on whether the system does it right. In the middle, prototyping and validation link design with implementation, helping to reduce risk early. At every step, system developers must assess cost, risk, and system value, providing insights that are fed back into earlier phases as needed. The Vee is not a one-way process; instead of following a fixed sequence, it allows for continuous refinement as new insights emerge. Figure 1.1 shows how this version captures both the structure and adaptability of the systems engineering process.

Using the Vee as an example, a critical element of abstraction is integrating systems mapping, which visually represents relationships and dependencies. Systems mapping is a tool that illustrates interconnections, feedback loops, and dependencies between system components. It helps identify causal relationships, bottlenecks, leverage points, and unintended consequences, making it easier to navigate complexity in decision-making. This approach ensures that decisions are made with a comprehensive understanding of their system-wide impact [58].

Applying system mapping tools helps promote a more holistic approach to system design by illustrating how various components interact and influence one another. Agent-based modeling (ABM) adds to this by offering a tool for simulating the interactions of individual components within a system [59].

At its core, ABM involves creating a virtual environment filled with autonomous agents, each with its own rules, behaviors, and attributes. By simulating interactions among these agents, agent-based modeling provides insights into system dynamics, aiding in the anticipation of potential outcomes that may not be immediately clear [60].

This approach enables the analysis and prediction of emergent behaviors, or unexpected outcomes, that arise from the complex interaction of individual system elements.

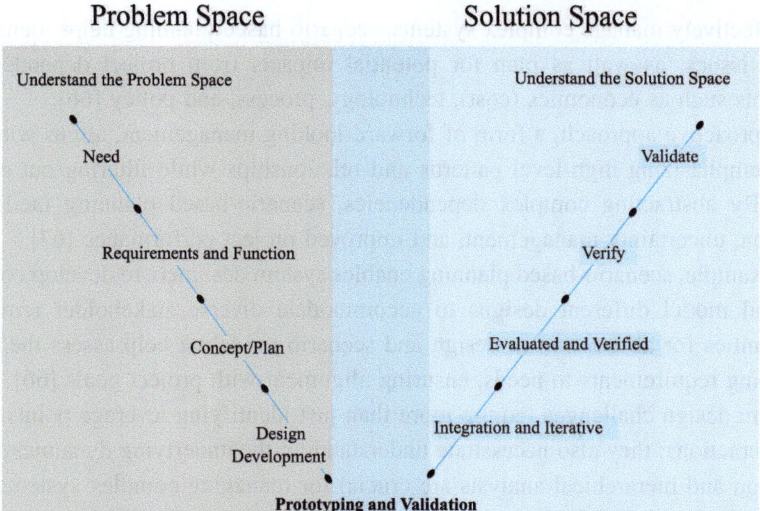

Fig. 1.1 A perspective on the systems engineering Vee

Abstraction enables system designers to focus on high-level patterns and interactions rather than getting lost in the details of specific causes. By abstracting certain behaviors and rules, ABM helps reveal broader system dynamics, making it easier to analyze trends, identify leverage points, and predict outcomes in complex systems [61].

While ABM offers significant advantages, it also comes with challenges. A primary concern is the computational intensity required to simulate large numbers of agents and their interactions. As the complexity of the model increases, so does the processing power and time needed for accurate simulations, potentially limiting scalability and efficiency [62].

Another challenge is the complexity of validating and calibrating agent-based models against real-world data [63]. Understanding linear models is only the starting point in complex systems. Nonlinear dynamics, where effects are not directly proportional to their causes, play a significant role and must be understood to prevent potential system failures. Abstraction helps manage this complexity by allowing system designers to focus on high-level patterns and interactions rather than getting lost in every detail.

Even small changes in input variables can trigger significantly large and unforeseen consequences, a concept borrowed from chaos theory. Engineers can model these relationships more effectively by deliberately employing abstraction, identifying critical leverage points, and mitigating risks before they escalate [64].

Consequently, drawing from chaos theory to create an abstraction-based analysis, it is clear that seemingly random systems adhere to specific rules; nevertheless, minor alterations can result in significant effects over time, complicating long-term forecasts [65].

To effectively manage complex systems, scenario-based planning helps identify trends and key issues, as well as plan for potential impacts from project dependencies and constraints such as economics (cost), technology, process, and policy [66].

This proactive approach, a form of forward-looking management, aligns with abstraction by emphasizing high-level patterns and relationships while filtering out extraneous details. By abstracting complex dependencies, scenario-based planning facilitates risk mitigation, uncertainty management, and improved project performance [67].

For example, scenario-based planning enables system designers to develop contingency plans and model different designs to accommodate diverse stakeholder requirements. Opportunities for abstraction in design and scenario modeling help assess the efficiency of mapping requirements to needs, ensuring alignment with project goals [66].

System design challenges require more than just identifying leverage points and modeling interactions; they also necessitate understanding the underlying dynamics. Adequate abstraction and hierarchical analysis are crucial for managing complex systems. In complex projects, evolving from requirements to functions to tasks, this approach benefits from breaking down detailed processes into multiple levels and visualizing relationships in a clear, hierarchical structure for better understanding and decision-making [68].

Abstraction enhances problem-solving by emphasizing core concepts and key properties while filtering out irrelevant details [65]. Hierarchical analysis, on the other hand, involves decomposing a system into structured levels to clarify the relationships between higher-level functions and lower-level components [68].

When organizing systems, developers must structure data in a way that is meaningful to reduce ambiguity and clarify requirements. Implementing hierarchical structures in design enhances modularity, providing a clearer understanding of system organization and functionality.

Abstraction can be introduced into project development through four fundamental questions: What? Who? When? and How? This framework serves as a core project management adage, offering a structured approach to decision-making [69]:

- What? Defines project goals, deliverables, and scope.
- Who? Identifies key stakeholders, roles, and responsibilities.
- When? Establishes timelines, milestones, and deadlines to ensure project completion.
- How? Determines processes, methodologies, and resources for execution.

Applying this structured approach creates layers of abstraction, making it easier to analyze, modify, and manage individual components without requiring complete knowledge of intricate details. Effective decision-making in engineering relies on experience and the lessons learned from past projects. This process involves "identifying and mitigating risks, improving decision-making, and enhancing overall project performance," ensuring that each iteration is stronger [70].

Abstraction plays a crucial role in simplifying complex systems, making them more understandable and practical. This is highlighted by the preference for straightforward explanations in decision-making contexts [71]. This simplification is essential for developers, enabling them to focus on key components and interactions, resulting in more efficient designs. Diagnostic techniques further enhance system planning by identifying risks, vulnerabilities, and inefficiencies, which are necessary for informed decision-making to aid in mitigating complexity and realizing a solution [71].

Continuing with the theme of simplicity, Claudia Sehl, lead author and a PhD candidate in developmental psychology at Waterloo, collaborated with professors Ori Friedman and Stephanie Denison to study how people prefer explanations and solutions. Through seven experiments involving 2820 participants, the research investigated whether individuals preferred more straightforward or complex explanations for an outcome or a goal. The results consistently preferred simpler options, particularly those based on common and reliable causes. When a cause was perceived as rare or unreliable, participants found it less helpful. The more familiar and dependable a cause was, the more appealing it was for understanding events and achieving future results [72].

This preference for simplicity and efficiency aligns with the principles of abstraction in system design and problem-solving. Abstraction simplifies complexity by focusing on essential elements while filtering out extraneous details, allowing for better decision-making and clarity. Just as people prefer straightforward explanations in everyday reasoning, system engineers and developers use abstraction to create models, frameworks, and structures, emphasizing critical functions without unnecessary complications. Sehl noted that fewer causes were seen as faster and more effective, suggesting a shared cognitive preference for abstraction in explanations and goal achievement.

Friedman emphasized that efficiency, or doing more with less, is a key factor in how people think about reasoning and problem-solving. Similarly, leveraging abstraction in engineering and project management enables the development of scalable, adaptable systems by focusing on high-level structures and essential relationships, thereby ensuring efficiency and effectiveness [72].

Ultimately, learning from each experience and continuously refining approaches to leverage abstraction effectively, understand nonlinear dynamics, and build resilient systems [73].

1.5 The Role of Abstraction in Simplifying Complexity

Abstraction, at its core, is about simplifying complex reality by focusing on essential characteristics while omitting irrelevant details. It allows system designers to view complex systems through a simplified lens, facilitating comprehension, modularity, and efficiency. By abstracting away implementation details, developers can concentrate on functionality without being burdened by intricate technicalities [74].

Abstraction enables system developers to filter out nonessential details during system design, allowing for a focus on the overall architecture. Breaking away from unnecessary information is essential, as abstraction manages complexity within the system design, allowing more significant emphasis on desired outcomes [75]

Through a high-level representation of underlying elements, developers can concentrate on functionality without being burdened by intricate implementation details. This is particularly evident in database management systems, where abstraction separates the logical and physical layers, allowing developers to work with data models without needing to understand the complexities of the storage engine. For example, a database abstraction layer serves as an intermediary between the application and the database, isolating developers from low-level storage intricacies. This abstraction enhances portability, optimizes performance, and improves code maintainability. By abstracting implementation details, system designers can focus on higher-level functionalities, encourage code reusability, and develop robust, flexible systems that adapt to evolving requirements [35].

Abstraction aids in managing complexity in distributed systems by identifying how to organize interconnected components that must operate across multiple machines and communicate over networks. Abstraction simplifies coordination, scalability, and reliability in distributed systems by hiding low-level complexities and providing standardized interfaces for interaction. Consider the following [35]:

- Coordination: Abstraction enables seamless communication between distributed components by providing middleware layers, such as message queues or service registries. This confirms components interact efficiently without needing to manage underlying networking complexities.
- Scalability: By abstracting resource allocation and workload distribution, load balancers and automated deployment platforms (e.g., Kubernetes) enable systems to scale dynamically without requiring developers to manually manage infrastructure details.
- Reliability: Abstraction encapsulates fault tolerance mechanisms, such as data replication, failover strategies, and automated recovery, ensuring that failures in individual components do not disrupt the entire system.

By structuring system interactions and hiding implementation complexities, abstraction enables developers to focus on higher-level design and functionality, resulting in more efficient, scalable, and resilient distributed systems. An example of systems integration is the message queue, which abstracts communication between components, allowing for asynchronous message exchange. This simplifies synchronization, enhances fault tolerance, and enables load balancing. Another essential abstraction is the distributed file system, which provides a unified storage view across multiple machines, facilitating horizontal scaling while maintaining fault tolerance and data consistency. Techniques such as replication, sharding (splitting a database into smaller, more manageable pieces), and load balancing serve as abstraction techniques, hiding the complexities of failure recovery while ensuring system availability and reliability [35].

Abstraction isn't a one-size-fits-all concept; it operates at different levels and layers, each serving a distinct purpose. Abstractions are created for a particular purpose as problem-solving tools designed to address a specific problem, challenge, or goal [76].

At a high level, a system can be described by its overall functionality, while lower levels focus on specific implementation details. For example, in a software application, the user interface serves as a high-level abstraction, concealing the underlying code and data structures, allowing users to interact with the system without needing to understand its internal workings [77].

Code can be structured into layers of abstraction, with each layer providing services to the one above it. This layered approach simplifies development and maintenance by enabling developers to focus on specific system components without needing to grasp all underlying intricacies [78].

This approach also provides the benefit of enabling different systems to work together without requiring an understanding of the underlying intricacies [77]. While abstraction is a powerful tool, finding the right balance is crucial. Excessive abstraction can lead to over-engineering, making the system more challenging to understand and maintain. This phenomenon, often referred to as "abstraction hell," occurs when excessive layers of abstraction obscure rather than clarify the system [77].

As Joel Spolsky notes, "All non-trivial abstractions, to some degree, are leaky," meaning that underlying complexity can still surface, causing unexpected issues. Personal experience often reinforces this reality, as developers encounter situations where abstractions fail to fully shield them from the system's deeper intricacies [79]. Over-simplifying abstraction is like simplifying a recipe to the point where essential steps are missing, resulting in failure. To avoid this, abstractions must strike a balance between simplification and necessary detail to maintain system integrity.

Additionally, effective dependency management is essential, ensuring that relationships between components are clearly understood and properly maintained. Managing causal dependencies requires a systematic approach that accounts for both direct and indirect relationships between elements.

Addressing these dependencies early in the development process helps prevent bottlenecks and unforeseen complications. A well-designed abstraction should simplify the system without sacrificing critical details or introducing unnecessary complexity. This ensures that the system remains manageable, maintainable, and adaptable to future changes [78].

However, it is important to recognize that "not all abstraction is good". Poorly structured abstractions can introduce more complexity and confusion, making a well-thought-out and clearly defined approach essential [74].

1.6 Hierarchical Analysis: Structuring Understanding

Hierarchical analysis complements abstraction by providing a systematic approach to simplify system analysis and design, enabling an understanding of how a system is organized, such as structuring into distinct, ordered layers with specific responsibilities [80].

The hierarchical metaphor describes the layered structure associated with planning and managing overall project development. This structure organizes tasks, objectives, and decisions in a layered manner, with higher levels representing broader goals and lower levels detailing specific actions required to achieve them [68]. Change networks or stakeholder maps are often used to define the flow of information from one element to another, organized to meet a specific purpose or illustrate stakeholder interactions. The ability to represent these flows is what makes such diagrams valuable in managing complexity and communication [81].

Within the systems engineering V-model, this same principle applies when mapping functional or interface relationships. An N2 diagram offers a structured way to capture and analyze interactions between system elements, making it a powerful tool for visualizing integration points and dependencies across the Vs' development and verification paths [81].

For example, translating the systems engineering Vee into a hierarchical structure, instead of relying solely on the classic V shape, provides a clearer view of layered dependencies throughout the systems' lifecycle. One key benefit of this approach is the ability to represent the flow of information from one element to another, organized as necessary to reflect the system's complexity and interactions.

This hierarchical perspective functions like a controlled waterfall, where each phase informs the next, guiding design details as development progresses. The structure resembles an N2 diagram by emphasizing interfaces and interactions between components, reinforcing the value of a layered view. These relationships help drive design decisions by revealing interdependencies among system elements, ultimately supporting a more coherent and integrated development process.

To highlight the relational descriptions, the cells in Table 1.1 are marked with "O" to denote outputs produced by a phase and sent to others, "I" to indicate inputs received from earlier stages, and "X" where a phase references its internal activities or artifacts. This format establishes a pattern for how information flows across the system lifecycle. Every phase is contextually linked, ensuring that no activity stands alone. Each decision is informed by prior work and contributes to subsequent phases, reinforcing both traceability and system coherence.

A hierarchical view of the Systems Engineering Lifecycle depicted by the Vee breaks down each phase into defined objectives, supporting activities, and resulting system interactions. Abstraction levels are introduced to manage complexity, maintain architectural clarity, and ensure alignment across domains. Building on this structured approach, N2 diagrams offer a complementary perspective by focusing on the interactions between system elements. While the hierarchical view emphasizes what happens within each phase, the N2 diagram highlights how components communicate across phases, making this a critical tool for tracing interfaces, managing dependencies, and supporting integration throughout the lifecycle.

Table 1.2 presents a hierarchical view of the Systems Engineering Lifecycle, breaking each phase into clear objectives, supporting activities, and resulting system interactions. Abstraction levels are used to manage complexity, preserve architectural coherence, and keep efforts aligned across different domains.

This structured approach is useful for navigating the complexities of systems developers. By breaking down each phase into clear components such as objectives, actions, and interactions, engineers can maintain focus and ensure that every element supports the system's overall integrity. Abstraction plays a critical role by isolating concerns at appropriate levels, which eases mental workload and enhances analytical precision.

The table enhances traceability by supporting requirement verification, process management, and stakeholder alignment. A common framework facilitates integration across

Table 1.1 N2 diagram with interfaces and interactions [87]

	Identify needs	Requirements	Concept	Plan/Design	Develop	Test	Verify	Acceptable	Validate	Operate	Retire/ Dispose
Identify needs	X	O	O	O	O	O	O	O	O	O	O
Requirements	I	X	O	O	O	O	O	O	O	O	O
Concept	I	I	X	O	O	O	O	O	O	O	O
Plan/Design	I	I	I	X	O	O	O	O	O	O	O
Develop	I	I	I	I	X	O	O	O	O	O	O
Test	I	I	I	I	I	X	O	O	O	O	O
Verify	I	I	I	I	I	I	X	O	O	O	O
Acceptable	I	I	I	I	I	I	I	X	O	O	O
Validate	I	I	I	I	I	I	I	I	X	O	O
Operate	I	I	I	I	I	I	I	I	I	X	O
Retire/ Dispose	I	I	I	I	I	I	I	I	I	I	X

Table 1.2 Hierarchal view of systems engineering lifecycle [87]

Phase	Purpose	Support activity	Result	Abstraction level	Focus
Identify needs	Understand the problem and the opportunity to execute a project	Stakeholder needs and requirements gathering. Capability assessment	Needs statements, goals and objectives	Problem space	What needs to be done and why it is important
Requirements	Determine what the system development do	Derive baseline system and stakeholder requirements and document	Requirements specifications	Operational context, systems level	Determine what the systems must do
Concept	Identify viable approaches to satisfy requirements such as potential configurations, architectures or frameworks	Trade studies, models and simulations, feasibility analysis	Concept of operations	Design conceptualization architectural focus	How the system could be structured. Trade studies, analysis of alternatives
Plan/Design	Plan the development that will support the design of the system	System architecture, interface design, risk assessment. Determination of constraints and dependencies	Systems engineering plans, project management plans, schedules, work breakdown structures, cost estimates	Logical/Physical analysis	How the system will be built. System and interface documentation
Develop	Build system components, software and interfaces	Code software, build or fabricate and integration of subsystems to realize whole system	Completed components or prototypes	Subsystem/ component level, whole system analysis	The realization of the actual systems elements. Code, hardware and subsystem builds
Test	Check individual components or subsystems	Assessing system performance and behavior under simulated operational conditions using unit and bench testing	Test plans and reports	Unit element, subsystem	Does each system element work as expected
Verify	Ensure the system was built as per specifications	Requirements traceability	Evidence based verification tracking	Integration level	Do assembled systems elements satisfy specifications

(continued)

Table 1.2 (continued)

Phase	Purpose	Support activity	Result	Abstraction level	Focus
Acceptance	Ensure the system is ready to be delivered to the stakeholder	Acceptance testing with stakeholder in the operational setting	Performance reports necessary to facilitate acceptance approval	Readiness check	Is the system development complete and conform to stakeholder expectations
Validate	Ensure the right was built and meets needs and requirements	Operational test	System validation analysis documented and concurred by stakeholder	User-level validation	Does the system meet actual operational needs
Operate	The system developed operates as intended	Training, lifecycle support to ensure proper operation	Operational data, failure and usage data	Real-world operation	Is the systems working under real-world conditions, will updates or upgrades be needed?
Retire or dispose	System has reached end-of-life. System decommissioning will need to occur safely, securely and responsibly	System closeout, documentation archiving, and hazard/security compliance checks	System retirement, hazardous material disposed and final results completed	System end-of-life	What special considerations will needed to properly address how to retire the system

disciplines, making dependencies and trade-offs more visible and manageable. By organizing development into clear phases with defined outcomes, teams can collaborate more effectively, reduce rework, and maintain focus on building reliable systems. The result is a lifecycle that remains cohesive, traceable, and purpose-driven from concept to retirement.

Instead of serving as a simple timeline, the hierarchical model demonstrates how each phase influences the next. The feedback loop it establishes ensures the system remains traceable from concept to evaluation. As a structural and procedural guide, the model outlines dependencies, supports iteration, and connects every step to the systems' broader goals.

Examining the Systems Engineering Vee Model through a hierarchical perspective reveals how it facilitates complexity management by organizing system development into structured, traceable layers. For instance, under the Plan/Design phase, Table 1.2 references the Work Breakdown Structure (WBS). When integrated with a Work Breakdown Structure (WBS), the analysis of planning and development evolves from a simple process map into a framework that aligns engineering tasks directly with the system architecture. The analysis of high-level requirements is decomposed into subsystems and components, and the structure closely mirrors that of the WBS. Each WBS element defines a specific scope of work tied to a corresponding part of the system.

In a Work Breakdown Structure (WBS) diagram designed to reflect hierarchical analysis, each box represents a level of decomposition from general to specific. The example below illustrates the first two phases of this structure as outlined in Table 1.2, where needs and requirements are broken down to initiate the WBS.

The work breakdown structure (WBS) is a valuable tool for breaking down a project's scope into smaller, manageable deliverables or tasks, organized in a hierarchical, tree-like structure. While there are various types of WBSs, the key advantage is in understanding the project's structure and dependencies [82]. Consider the following (Fig. 1.2) [82]:

- Deliverable-Based WBS: Focuses on the deliverables or outcomes of the project.
- Phase-Based WBS: Breaks down the project into distinct phases or stages.
- Responsibility-Based WBS: Assigns tasks and responsibilities to specific team members or needed resources.

As the project progresses, the hierarchical structure accounts for integration, verification, and validation by embedding these concerns into defined dependencies across the system. Each element in the Work Breakdown Structure (WBS) is linked to specific requirements, ensuring components are not only developed according to plan but also verified against their intended purpose. This alignment reinforces traceability, strengthens accountability, and provides a structured path from initial concept to fully validated deliverable. By exposing relational dependencies, the hierarchy reduces ambiguity, clarifies responsibilities, and supports a clear understanding of what needs to be delivered, when, and by whom.

For example, in designing a software system, a hierarchical approach might involve first defining the overall system goals, then breaking down those goals into smaller subgoals, and finally specifying the tasks and functions required to achieve each sub-goal.

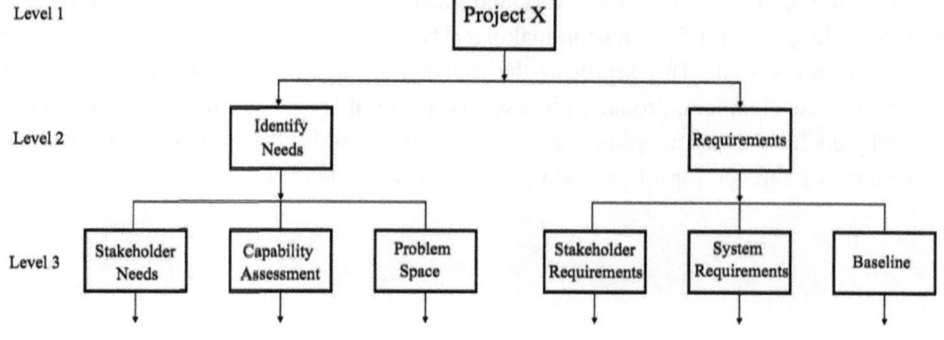

Fig. 1.2 WBS mapped from hierarchical view [88]

This structured approach ensures that all aspects of the system are considered and that the relationships between different components are well defined. In this way, engineering abstraction identifies what element can be used to address a specific function (technical) or task (programmatic). Understanding relationships, dependencies, and potential constraints helps identify reusable components and improves design efficiency by revealing how these factors may impact the overall project or system development [68].

1.7 Using Abstraction Hierarchies to Manage System Complexity

In systems engineering, a framework is a structured approach to gathering information needed to plan, develop, and manage complex systems. It provides a consistent set of processes, tools, and guidelines that help project development stay aligned, help manage risk, and define quality and efficiency parameters throughout the system's lifecycle.

Such a framework helps organize development tasks to support decision-making, clarify roles, and ensure all essential activities, such as requirements definition, design, integration, and testing, are systematically aligned and addressed. Frameworks typically include:[83]

- Defined processes and development phases: Defines the relevant sequence of activities from project concept to deployment, supporting a systematic approach to system development [83].
- Best practices and standards: Implement a systematic and structured approach to development by utilizing industry standards, such as ISO/IEC/IEEE 15288, to enhance consistency and reliability [83].
- Modeling and analysis tools: Modeling is the process of simplifying complex real-world systems to improve understanding, design, and management. It uses graphical, mathematical, or physical representations to abstract reality and remove unnecessary complexity. These models help visualize, simulate, and analyze systems, making it easier to comprehend how they work and identify ways to optimize performance [84].
- Establishing criteria for decision-making: The goal is to outline the functional baseline for the system. This serves as the foundation for creating detailed specifications (such as development, product, process, and material specifications) by the designated designers for the various components of the system. Additionally, it acts as the primary technical reference for all program planning documents [85].

1.8 Translating Declarative and Procedural Knowledge into Practice

Both declarative and procedural knowledge are essential in complex engineering environments, supporting both understanding and effective action [38]. Translating this knowledge type into practical application involves the following:

- *Understanding System Context*: Identify key focus areas, such as stakeholder needs and system capabilities, while recognizing domain-specific challenges and constraints. For example, when designing a wireless communication system, a key stakeholder might require broad geographic coverage, while a critical system capability could be maintaining signal reliability under high network traffic [86].
- *Applying Engineering Techniques*: Extract key information such as costs, schedules, and technical performance parameters. Run models, perform analyses, and carry out verifications to ensure system objectives are met. For example, during the planning phase of an autonomous vehicle project, engineers estimate development costs, establish delivery timelines, and define performance metrics such as sensor accuracy, processing speed, and system response time under various driving conditions [87].
- *Iterating and Refining Continuously*: Evaluate and adjust designs based on feedback and evolving requirements. This step involves collecting relevant data to assist in decision-making and development. For example, in designing a new drone navigation system, engineers gather information on battery life, payload capacity, environmental constraints, and regulatory requirements. This data supports trade-off analysis for assessing design alternatives [88].

1.8.1 Integrating Declarative and Procedural Knowledge in System Development

Combining declarative and procedural knowledge provides developers with actionable insights that guide decision-making, enhance adaptability, reduce risks, and strengthen collaboration. This integration ensures that critical information is accessible and effectively supports development. Effective project development necessitates the integration of both practical skills and theoretical understanding. This combination is crucial for addressing complex challenges, improving adaptability, and driving successful engineering and system design outcomes. By merging hands-on expertise with core concepts, development teams can navigate complexities and optimize project execution. To apply both approaches, consider the following [89]:

1. Combining Practical Skills with Theoretical Understanding

Effective project development relies on both practical skills and a deep theoretical understanding. This balance is crucial when managing new or complex projects, where the margin for error is minimal and adaptability is required. Consider the following example:

Consider a fast-food restaurant aiming to implement an automated ordering system to address common pain points—long wait times, order errors, and inconsistent staff interactions. Insights from restaurant managers, counter staff, and customers contribute to shaping practical, actionable solutions. These perspectives help identify bottlenecks, miscommunication issues, and variations in customer preferences. Additionally, understanding technical concepts such as user experience (UX) design, real-time payment processing, and order queue management ensures the system is tailored to resolve these challenges effectively.

2. Declarative and Procedural Knowledge

Declarative and procedural knowledge are two core types of understanding that support effective systems engineering. Declarative knowledge involves knowing facts and information, such as system requirements, design standards, or component functions. Procedural knowledge refers to the ability to perform tasks effectively, such as running tests, building models, or troubleshooting issues. Both are essential for connecting technical work to real project needs. When combined with sound engineering practices, they enable teams to make informed decisions, meet stakeholder expectations, and improve overall system performance. Integrating knowledge with technical principles enables the project to fulfill stakeholder needs and improve overall operations. Consider the following example [87, 89]:

Declarative Knowledge (The "What")

Declarative knowledge must provide clarity on key concepts and domain-specific insights that shape understanding:

- Customer pain points include operational issues such as excessive wait times, inaccurate orders, and inconsistent service quality.
- Operational feedback from managers, staff, and customers provides critical data for identifying inefficiencies and informing targeted improvements.
- Core challenges involve pinpointing process bottlenecks, resolving communication gaps, and addressing diverse customer expectations.
- Technical considerations include areas like user interface design, secure and efficient payment systems, and effective queue and flow management.
- System integration focuses on aligning day-to-day restaurant operations with digital tools and technologies to improve performance and customer satisfaction.

Procedural Knowledge (The "How")

Procedural knowledge defines the methods and processes to apply understanding effectively:

- Effective Issue Resolution: Engineering system solutions that reduce wait times and minimize operational errors.
- Solution Development: Converting stakeholder insights into practical, scalable system functionalities.
- Operational Improvement: Structuring system design and implementation to enhance customer satisfaction and optimize workflow efficiency.

3. Integrating Process and Purpose: Combining process with purpose ensures alignment with stakeholder needs and optimizes resource utilization. For example, in the automated ordering system:
   - Process (How its' done): Self-service kiosks enable customers to place orders, verify availability, and process payments. The system integrates with inventory management to ensure ingredient availability and prevent shortages.
   - Purpose (Why its' done): Streamlined ordering minimizes miscommunication, reduces delays, and improves the customer experience while optimizing operations.
4. Balancing Action with Insight: This approach merges action (execution) with insight (understanding) to drive meaningful results. Acting without insight risks errors, while insight without action leads to stagnation. A balanced approach ensures that decisions are informed, adaptable, and effective.
   Example:
   - Action (Execution): Self-service kiosks are installed to handle orders, process payments, and communicate with the kitchen, ensuring smooth operations.
   - Insight (Understanding): Before implementation, feedback, order patterns, and operational challenges were analyzed. This data informed the system's requirements, optimizing communication, order flow, and alignment with customer expectations.
5. Connecting Methods with Meaning: Understanding the rationale behind methods clarifies why they are used, how they function, and when they are most effective.
   Example:
   - Method (Technique Used): The system features real-time order tracking, enabling customers to monitor order status through digital screens.
   - Meaning (Rationale Behind It): This feature reduces customer anxiety, minimizes inquiries, and enhances service efficiency by keeping customers informed.
6. Unifying Practice and Principle: Effective development aligns practical applications with foundational concepts, standards, and best practices that guide system design.

These include modularity, scalability, and user-centered design, which assist decision-making, manage alternatives, and ensure consistency.
Example:

- Practice (Application): The system employs self-service kiosks to streamline order-taking, reducing dependency on counter staff.
- Principle (Foundational Concept): This principle follows the principles of efficiency and user autonomy, empowering customers to manage their orders, optimize operations, and enhance service flow.

These example illustrates the kind of analysis required to structure a project effectively. It highlights the practical "how" and the conceptual "what," ensuring system development is methodical and purpose-driven. This initial analysis occurs before technical development begins and directly influences key design and implementation decisions.

Systems engineers refine their understanding by striking a balance between practical execution and conceptual clarity, enabling them to plan, develop, and integrate solutions that meet the requirements of functionality, efficiency, and user experience [37].

The key challenge is to avoid unnecessary complexity while gaining meaningful insights. Achieving this balance relies on structured methods, such as prototyping, testing, modeling, and abstraction, which help translate high-level concepts into functional and effective systems [37].

Developing complex systems, such as an automated restaurant ordering kiosk, rarely results in perfect functionality right from the initial implementation. It usually goes through several iterations before achieving the desired performance standards. This iterative process involves extensive testing and user interaction to enhance its performance and identify usability issues. Additionally, software and subsystems must be analyzed to determine necessary changes, modifications, or optimizations for successful deployment.

Managing this level of complexity while gaining a deeper understanding relies heavily on how effectively information is derived, structured, and organized. Breaking down key elements into manageable components is essential to avoid cognitive overload. Simultaneously, maintaining a structured approach promotes continuous learning, efficient data management, and logical organization of details. Consider the following development [37]:

1. Defining Clear Objectives: Establish what success looks like and identify the most critical outcomes. Clear objectives provide a reference point for development and evaluation.
2. Identifying Critical Components: Focus on the subsystems or features with the highest impact or risk. This prioritization allows for targeted testing and improvement.
3. Sequencing Development Tasks: Organize tasks in a logical order that facilitates iterative testing, feedback collection, and refinement. Proper sequencing prevents bottlenecks and enhances productivity.

4. Tracking Dependencies and Constraints: Understand the factors that affect each component and their interrelations. Clear mapping of dependencies helps anticipate challenges and manage risks.

Establishing the right balance between abstraction and detail is essential for effective system development. Testing, prototyping, and structured abstraction help highlight important information while maintaining a focus on broader objectives. While these methods may require additional time and resources, they are instrumental in identifying design gaps and validating functionality.

Modeling and structured abstraction also help visualize component operations across different levels of the system architecture. At the highest level, a conceptual framework defines the systems' purpose and goals, offering stakeholders a comprehensive and easily understandable view. This organized methodology helps pinpoint high-level objectives and critical technical details necessary for developing the required solution. A structured, iterative approach enhances clarity and understanding, driving successful implementation and ensuring that complex systems are functional and aligned with stakeholder expectations [50].

In system development, abstraction is applied through a hierarchical approach that integrates top-down design with bottom-up implementation. At the highest level, systems engineers define functional objectives and abstract key functions to determine the most effective ways to meet those goals. The middle layer comprises generalized functions and reusable building blocks that translate abstract concepts into actionable components. At the lowest level, the focus shifts to physical implementation details and specific technical solutions. This layered structure ensures clarity, consistency, and traceability throughout the development process [50].

This analysis is necessary during project development, as each progressing phase requires new insights and information for planning and conceptualization. Abstraction, modeling, testing, and prototyping are utilized to explore solutions and uncover critical answers. While prototyping and testing demand time, resources, and planning, abstraction and modeling also have similar demands. However, these methods inherently support the iterative process of learning and adaptation, which can improve both the development experience and the quality of the final system.

Designing engineering systems that are efficient, effective, safe, and reliable necessitates the coordinated integration of multiple disciplines, including structured modeling, functional and contextual modeling, data management, data science, artificial intelligence, and digital twin technologies. While each area is well-established individually, comprehending how they interact to form adaptive and complex systems that evolve remains a challenge [90].

Abstractions must strike a balance between ambiguous generalization and unnecessary complexity. This is needed to focus the engineering processes without getting lost in unnecessary details; abstractions can be framed in terms of their level, scope, and goals, such as:

- Enterprise level: Abstractions focus on business processes, organizational models, and strategic objectives.
- System level: Abstractions represent the functional structure, showing how significant components interact to fulfill system goals.
- Application level: Abstractions are applied across system design, guiding structure, behavior, and interface logic in both software and hardware components.

Abstraction is a fundamental concept in computer science and an essential tool in systems engineering for managing complexity in software development and system design. It enables developers to focus on relevant details while concealing underlying complexities, simplifying both code and system structures. By filtering out unnecessary information, abstraction helps reduce ambiguity, clarify system behavior, and support consistent communication among stakeholders. Beyond planning and implementation, it also improves efficiency by enabling deeper analysis throughout the development lifecycle, including opportunities for better testing and refinement [90].

Although often viewed as a means to simplify complexity, abstraction can also be understood as a tangible structure in which individual elements are separated, examined, and refined. It serves as a conceptual filter and a practical framework for managing an analysis function in building effective, well-structured systems. To clarify the purpose and application of abstraction, consider the following definitions:

- An abstraction represents a concept of interest within a specific context. Each abstraction has a defined name, purpose, or intended function [90].
- Abstraction is a simplified representation of a system, used to support understanding, communication, clarification, or the design of specific system elements [91].

Abstraction is a simplified representation of a concept or system element within a specific context. It facilitates understanding, communication, and design by emphasizing relevant aspects and omitting unnecessary details. Each abstraction is defined by its name, purpose, and role, contributing to the structured and manageable development of complex systems [92].

1.9 Borrowing from Abstraction Engineering

The term "Abstraction Engineering" refers to the systematic practice of creating and applying abstractions to manage complexity in engineering systems. This approach involves developing simplified representations of complex systems that allow engineers to focus on essential aspects while concealing underlying details. By doing so, abstraction facilitates understanding, communication, and design across various system levels [90].

In software engineering, abstraction is essential for reducing complexity and enhancing efficiency. By enabling a focus on the key aspects of an object while hiding irrelevant details, abstraction streamlines both code and system architecture. Managing complexity becomes more straightforward by emphasizing critical features and excluding unnecessary details. This method, often referred to as information hiding, enables designers to focus on the most crucial elements, thereby making complex systems easier to comprehend, analyze, and construct [93].

1.9.1 System Development and Operational Context

The system under development should be purposefully designed to meet specific needs, establish new capabilities, or enhance existing ones. Effective planning is critical to successful system development, enabling a structured representation of how the system will function within its operational environment. This planning phase could incorporate a range of modeling techniques that help define, visualize, and optimize various aspects of the system and would include [90]:

- Structural Models: Represent the system's configuration and layout, highlighting key components and their interconnections.
- Process Models: Illustrate the flow of operations and interactions within the system.
- Historical Data Analysis: Use past data to forecast demand, identify patterns, and guide development.
- Predictive Models: Evaluate the potential impacts of performance, processes, or risks by simulating scenarios to identify risks and optimize outcomes.

Abstraction should not imply processing the entire planning or development cycle through an abstraction engine. Instead, it should focus on identifying the elements that benefit from abstraction and selecting the most effective abstractions to capture functional and non-functional system requirements. When modeling the functionality of an Artificial Intelligence (AI) and Machine Learning (ML) algorithm, abstractions can represent the algorithm's expected behavior as it learns and adapts. This enables analysis to verify that the system behaves as intended and allows tracking of behavioral changes based on state observations. In AI/ML learning, abstractions should also facilitate the integration of explicit data, such as a calendar, into training and retraining cycles, enabling models to incorporate new information as it becomes available [90].

Developing an abstraction is typically the role of a subject matter expert (SME) who understands the domain and its functionality. Requirements engineers, domain specialists, and quality assurance professionals may also contribute to the analysis. For example, when using a digital twin tool, SMEs outline the processes and identify the data to be collected. As mentioned earlier, some aspects of process behavior and

Table 1.3 Difference between bottom-up and top-down models [94, December 2]

Top-down design approach		
1	System definition	Identify the overall need, scope, purpose, and requirements defined
2	Functional decomposition	The system is divided into high-level subsystems and components
3	Specification development	Define the requirements, functions, and interfaces for each component or subsystem. At this stage, some initial performance criteria may be known, which can be utilized to derive additional requirements as needed
Bottom-up implementation approach		
1	Component development	Design and develop individual modules according to the requirements function
2	Subsystem integration	Integration involves combining different components or subsystems into a more extensive, complex system
3	System testing	Refers to confirming that a system, component, or function meets specified requirements by executing predefined procedures, making modifications as necessary, and analyzing results to determine whether changes are warranted or required

Retrieved from https://www.geeksforgeeks.org/difference-between-bottom-up

measurement relevance can be learned, allowing machine learning algorithms to help shape the abstractions, for instance, by incorporating a predictive model. When complex or interdisciplinary questions arise, or unexpected analyses are needed, users may need to define their system views. User-friendly tools that support creating such abstractions are especially valuable [90].

Abstractions can be conceptualized in a hierarchical, top-down manner or a bottom-up approach. In some applications, the fundamental abstractions encapsulate observable phenomena and necessitate the involvement of subject matter experts who employ a top-down methodology to develop the requisite models. A top-down design approach typically defines the hierarchical structure of a system's development.

The resulting framework is based on the high-level system goals and functions that were first described. This helps structure the bottom-up implementation, as individual subsystems or components are defined and integrated to realize the system as needed. Consider the Table 1.3:

The type of analysis often determines how abstractions are formulated, either through hybrid or adaptive strategies. In these strategies, an initial abstraction is defined by a human and then refined through a bottom-up approach using data. A typical example of this is the operational environment model used in autonomous system development.

Typically, experts design the language of these models, known as a meta-model. At the same time, the system utilizes sensor data and learning-augmented components to gradually populate the model with specific object instances, including pedestrians, traffic signals, road configurations, obstacles, and other vehicles.

Abstractions are created and discovered to support cognitive technologies, such as AI/ML or autonomous system development. Some abstractions are crafted with input from domain experts who identify which aspects of real-world applications are relevant to the cognitive system's intended functionality, ensuring these design elements are integrated into the design framework [90].

Integrating data sources with dynamic schemas and content, such as user-generated content on social media platforms, often reveals the need for iterative abstraction in personal assistant systems. These systems must continually analyze incoming data, refine their abstractions to reflect evolving user behavior, and adjust their responses accordingly. Through this ongoing process, AI-powered personal assistants can identify new patterns, enhance decision-making, and improve personalization to better serve the evolving needs of individual users [95].

Following an iterative approach is common in systems engineering to support refinement and adaptation throughout the development process. The initial concept of a project is rarely the final one. However, abstraction and hierarchical analysis help to break down a system into well-defined components while maintaining transparent relationships and dependencies, which makes updates and changes more manageable and less risky. The reduced risk is due to several factors:

- Modular Understanding: Systems are divided into smaller components, allowing for isolated testing and adjustments without impacting the entire system [96].
- Clear Dependencies: Hierarchical analysis clarifies the relationships between components, making it easier to understand and predict the effects of changes and avoid unintended consequences [97].
- Incremental Improvements: New features or modifications can be integrated smoothly without requiring a redesign of the entire architecture. Iterative development supports gradual enhancements and testing, allowing issues to be identified and addressed early before they escalate [98].
- Easier Troubleshooting: Well-abstracted components are easier to debug and optimize, preventing minor issues from escalating into major problems. Continuous testing and feedback loops help ensure that changes align with requirements and do not introduce new issues, maintaining system stability and reliability [99].
- Adaptive to Change: The approach enables the system to adapt smoothly to new requirements or environmental changes with minimal disruption. Systems designed with adaptability in mind can accommodate evolving requirements and technology shifts more efficiently [100].

1.10 Abstraction and Dependency Management in Systems Engineering

Systems engineering benefits from using abstraction and dependency management to define functionality, making system development less challenging. This approach identifies ways to reduce complexity-related ambiguity, helps manage dependencies, and increases the chances of meeting requirements. Integrating abstraction and dependency management into development improves adaptability and prepares systems for change and uncertainty. This approach enhances system reliability and supports continuous learning and improvement.

Abstraction reduces complexity by focusing on essential details, allowing developers to concentrate on key functionalities. Dependencies arise when system components rely on each other to function correctly. Managing these dependencies effectively is important for understanding. 'whole-system' analysis, especially when using abstraction. A focused approach is required for abstraction and dependency management as described below [101]:

- Dependency Direction: System developers can work with simplified models instead of dealing with ambiguous or unfamiliar details, facilitating the development process. In many systems, dependencies usually flow from low-level components to high-level ones. The conventional design is simple but often leads to tight coupling, making the system difficult to modify. A well-structured architecture addresses these challenges through dependency inversion, enhancing flexibility by reducing direct dependencies between high- and low-level components [102].
- Coupling: The degree of interdependence between components is known as coupling. This is important because it allows components to be isolated for focused analysis, making it easier to determine their functionality and dependencies with other elements [102].

  (a) Tight coupling occurs when components are highly dependent on each other, making modifications difficult.
  (b) Loose coupling allows components to interact through well-defined interfaces, enabling greater flexibility, easier scaling, and simplified maintenance.

- Interface Management: Clear, well-defined interfaces are essential for managing dependencies. Interfaces specify how components interact, what data is exchanged, and the communication protocols used. Proper interface management promotes loose coupling and enhances system adaptability [103].
- Abstraction Layers: In systems engineering, abstraction is typically organized into hierarchical layers, each providing specific services to the layer above it. This layered structure conceals the complexities of lower layers, streamlining development

and maintenance. Well-defined interfaces between these layers facilitate dependency management, ensuring clear communication and integration across system components [103].

- Dependency Injection: A design pattern that allows dependencies to be provided externally rather than being hardcoded within a component. This technique reduces coupling and makes testing more efficient [104].
- Procedural Knowledge: This area focuses on how processes, workflows, and methods manage dependencies. This includes selecting suitable design patterns, structuring code for modularity, and ensuring efficient communication between components [105].
- Declarative Knowledge: This type of knowledge focuses on the "what," providing an understanding of system architecture and the relationships between components. This aids in decision-making by offering a conceptual view of how dependencies influence the overall system [106].

Effective dependency management extends beyond rigid constraints, requiring a thorough understanding of the system's interconnections and their impact on overall functionality. Developers can architect and manage systems that adequately satisfy requirements by employing structured analysis and relational insights. This methodology fosters an environment of continuous learning and iterative refinement, facilitating the developers' progression in addressing ambiguity. Building transparent frameworks that efficiently manage interactions and dependencies ensures scalability, maintainability, and adaptability. This enables system development to evolve with changing demands while keeping complexity under control.

A common challenge in systems engineering is that, although each unique project faces distinct and complex issues, project developers often overlook the optimization of their problem-solving methods. Most analyses begin with a top-down systems engineering framework, regardless of the project's scope or stage in the engineering lifecycle.

This rigid mindset, often from a limited understanding of projects, leads to inefficiencies, repetitive cycles, and barriers to practical solutions. Instead of relying on a "one-size-fits-all" approach, project developers should recognize that different projects may require tailored strategies, be it top-down, bottom-up, or a hybrid model.

All of these approaches depend on the specific needs and requirements of the system. Understanding these interrelationships and recognizing the right time to apply the appropriate method is a fundamental competency in systems engineering. This skill directly affects project timelines, resource allocation, and overall success, thereby ensuring effective development and management of systems [107].

Causal dependencies are fundamental to the relationships within systems. In these dependencies, one component exerts a direct influence or dictates the behavior of another component. Such dependencies are inescapable, creating critical connections between the system's elements [108]. In software development, causal relationships manifest when one component's completion or functionality directly impacts another component's ability to operate effectively. This understanding helps identify critical paths and potential bottlenecks in system development [109] (Table 1.4).

Causality in system dependencies transcends mere correlation. Although correlation may imply a relationship between two elements, causality delineates a clear direction of influence. This differentiation is important for project managers and developers, as it enhances their understanding of the interrelations among components and how they affect each other. This comprehension facilitates the development of more effective risk management and resource allocation strategies. Moreover, understanding causal relationships empowers teams to more accurately predict and manage the effects of modifications within the system, allowing them to trace the ripple effects through interconnected components [110].

Table 1.4 Management dependencies

No	Term	Concept	Rationale
1	Modularity [111]	Breaking down a system into smaller, self-contained modules with well-defined interfaces helps manage dependencies	If dependencies are clearly defined and bounded, changes within a module are less likely to impact other modules. This supports loose coupling and makes the system more adaptable to change
2	Dependency [112] Management	Managing dependencies means understanding how distinct parts of a system connect, avoiding circular dependencies (where modules depend on each other in a loop), and using techniques like dependency injection to keep components flexible and independent	This helps avoid a chain reaction of changes when one module is updated, making the system more stable and easier to maintain
3–4	Scalability and Adaptability [113, 114]	A well-structured, modular system with managed dependencies is more scalable. New modules can be added without significant disruption to existing functionality	The system is also more adaptable to changing requirements, as modifications can be made within specific modules without affecting the entire system
5	System integrity [115]	By managing dependencies effectively, the system's integrity can be maintained	Changes in one part of the system are less likely to cause unintended consequences in other parts, leading to a more robust and reliable system
6	Causality [116]	Causality can be defined as the relationship between cause and effect. It asserts that every event has a cause or a set of causes that precede it and bring about its occurrence	This relationship forms the basis for reasoning, prediction, and explanation

Effective management of causal dependencies requires a structured approach that considers direct and indirect relationships among system components. Developers should identify and address these dependencies early to avoid issues. Utilizing tools and frameworks for understanding and monitoring can further improve efficiency and control [117]. Table 1.4 provides a summary of important considerations [111–117].

If the goal is simplicity, several factors must be considered. A recent study by Claudia Sehl, Ori Friedman, and Stephanie Denison, published in Cognitive Psychology, examined how people prefer simple versus complex explanations. In experiments with 2820 participants, people consistently chose straightforward answers and solutions. The study found that standard and reliable causes were more helpful, while rare or uncertain causes were less valuable. In other words, people favor clear and dependable reasoning to understand events and predict outcomes, highlighting the value of abstraction and simplifying complexity to make systems more understandable and practical [117].

Organizing information into a functional structure is necessary for systems engineering and project management. When planning a complex project, simplifying and breaking down all the intricacies and key elements helps lessen the ambiguity.

A structured approach can enhance our understanding by emphasizing core principles and patterns, thereby reducing cognitive load and fostering a more comprehensive understanding of complex problems. However, the challenge lies in achieving the right balance. After all, management is about making decisions [118].

1.11 Balanced Abstraction in Systems Engineering

Practical systems engineering demands a balanced approach to abstraction. Developers need flexible heuristics that evolve with experience and emerging best practices to enhance problem-solving and design scalability. Heuristics are the guiding principles in systems engineering that help navigate design decisions. Heuristics shouldn't be seen as static; they are dynamic, evolving as developers gain new experiences and encounter new challenges [119].

Adaptive heuristics help developers handle emerging technologies, shifting requirements, and unexpected complexities. Mastering abstraction in systems engineering requires continuous learning and a commitment to iterative improvement.

Developers must leverage experience and emerging insights to fine-tune their approach, ensuring that systems remain relevant as they evolve in development [120].

The challenge is finding the right level of abstraction to manage complexity. Too much abstraction can oversimplify important details, causing misunderstandings and mismatched requirements. Too little abstraction can overwhelm developers with unnecessary information, slowing progress and limiting flexibility. Consider the following examples of abstraction levels [121]:

- Too High: When designing a software system and abstracting the database layer, an overly high-level description might state, "data is stored and retrieved."
- While concise, this abstraction omits critical information about data formats, storage mechanisms, and access protocols. This lack of detail complicates implementation, leaving developers uncertain about how to interact with the data, potentially resulting in integration issues.
- Too Low: On the other hand, specifying every database query and table schema at the outset introduces excessive detail. Developers may become bogged down in low-level specifics, losing sight of broader system architecture. This granular approach hinders adaptability, making the system rigid and challenging to adjust to evolving requirements.
- Just Right: An optimal abstraction would represent the database as a service with well-defined interfaces for data access. This would hide internal complexities while providing developers with the information necessary for effective interaction. They would understand the available data and how to access it without managing backend specifics.

Finding the right abstraction level enables engineers to build scalable models and accurately represent the system's critical attributes. A well-structured analysis supports adaptability, allowing solutions to remain effective as project requirements evolve. Developers enhance efficiency and flexibility by systematically managing information and aligning it with analytical needs.

1.11.1 Real-World Application: Autonomous Vehicle Systems

A practical example of effective abstraction management is found in the design of autonomous vehicle systems. High-level abstractions handle decision-making tasks, such as route planning, while low-level abstractions manage precise controls, including steering and braking. At the higher level, roads are represented as nodes and edges in a graph, simplifying navigation. However, this abstraction can overlook real-world details, such as traffic lights, obstacles, or pedestrians. Bridging this gap involves integrating detailed models that include sensor data, vehicle dynamics, and real-time conditions. Properly managing these abstraction layers ensures the system remains adaptable and efficient, capturing both strategic objectives and critical technical details [118].

In systems engineering, managing complexity is an ongoing challenge. A practical approach to this is the abstraction hierarchy, a structured framework that breaks down complex systems into five progressively more detailed levels. It transitions from high-level goals to detailed physical implementations, clearly defining purpose and functionality at each stage.

In systems engineering, the abstraction hierarchy is a structured framework to understand and design complex systems, moving from high-level objectives to concrete implementations. The functional purpose is at the top of this hierarchy, which defines the system's overarching goals and objectives. This level establishes the foundational intent that drives the design and operation of all subsequent layers. Below it are the abstract functions, representing the essential capabilities required for the system to achieve its purpose. These functions focus on what the system must accomplish without detailing how it will be implemented. Progressing further, the general functions layer narrows the scope by outlining specific actions and behaviors necessary to support the abstract functions [122].

While still addressing what needs to happen, this level is more tangible and operationally focused. The next tier, physical functions, focuses on how the system will execute its intended purpose through defined processes and operations supporting the specified general functions. Finally, at the base of the hierarchy lies the physical form, which encompasses the hardware, materials, and components used to realize the system's capabilities. This structured decomposition from purpose to physical realization clarifies system design and supports practical analysis, troubleshooting, and optimization throughout the engineering lifecycle [122].

This layered structure is more than just a theoretical concept since it serves to deliver practical benefits throughout the lifecycle of system development. Engineers can better manage complexity by breaking down the system into understandable parts. Communication with stakeholders improves, as each level of the hierarchy provides a shared reference point for both technical teams and non-technical decision-makers. During the design phase, the hierarchy ensures the appropriate level of detail is introduced at the right time and in the proper context [122].

This structure also supports troubleshooting by allowing teams to trace issues from malfunctioning components back through the functional hierarchy. Modular design also benefits, as this approach makes it easier to reuse and adapt components across different systems or projects. In short, the abstraction hierarchy goes beyond organizing information and serves as a strategic method for designing, analyzing, and evolving complex systems with clarity and intent [122].

However, developers often introduce biases into this process. One common cognitive bias is functional fixedness, the tendency to view components or technologies solely in their traditional roles, which usually limits the potential for creative solutions. Popularized in broader discussions by author Dan Pink, this bias can cause developers to rely on familiar tools or past methods simply because they have been successful in the past. As mentioned earlier, experience is valuable. However, relying on familiar patterns without reassessing their relevance can stifle innovation and hinder adaptability. To avoid this, developers must engage in thoughtful and objective analysis that prioritizes actual system requirements over personal habits and preferences [123].

Hierarchical abstraction takes a broader approach to managing complexity by structuring information into layered groupings. Unlike the fixed structure of an abstraction hierarchy, hierarchical abstraction involves forming mental models that help individuals group similar elements and filter out unnecessary details. Complex projects require extensive amounts of information to manage development. To establish a clear understanding of what needs to be accomplished, this strains our limited cognitive capacity. However, these groupings allow us to compartmentalize information, simplify our thinking, and generalize without being overwhelmed by specifics. This approach is widely applicable across various domains and disciplines.

In practice, hierarchical abstraction is a valuable tool for supporting clearer thinking, better organization, and improved decision-making. However, applying it effectively also necessitates a small amount of reflection on previous experiences. Past successes and failures offer valuable insights, enabling more informed choices in future work. With expertise, system developers can refine their intuition, make more informed decisions, and achieve better outcomes [124].

The distinction between abstraction hierarchy and hierarchical abstraction goes beyond semantics. It highlights two fundamentally different approaches to organizing and interpreting complex systems. The key difference lies in their purpose and how they handle constraints. Abstraction hierarchy and hierarchical abstraction are not semantically different, as they represent two fundamentally different approaches to organizing and understanding complex systems.

The main difference between the two lies in their objectives (goals or purposes) and how they address the limitations of systems. The abstraction hierarchy illustrates how various levels within a system contribute to its overall functionality. This maintains a transparent and traceable connection between high-level functional objectives and low-level physical execution. This framework enhances analytical and diagnostic understanding, helping stakeholders, especially system operators and engineers, navigate the system's function and architecture across different levels of abstraction [50].

Hierarchical abstraction enhances cognitive efficiency by allowing system developers to conceptually group related components, which creates a streamlined and manageable mental framework. This abstraction facilitates faster and more generalized reasoning, aiding decision-making by filtering out unnecessary details. While both methodologies are valuable, each serves a different purpose. To utilize them effectively, it is critical to clearly define their scope and intended applications by selecting the right approach; whether analytical, operational, or conceptual, complexity can be managed without sacrificing clarity, usability, or functionality [50].

Understanding the differences between the abstraction hierarchy and hierarchical abstraction has essential implications for system design and management. Systems engineers can refine their understanding by balancing these elements, enabling them to plan, develop, and integrate solutions that meet the requirements of functionality, efficiency, and user interaction. When designing interfaces or troubleshooting systems, considering the abstraction hierarchy can help create more effective operator interfaces that support natural problem-solving patterns.

Sometimes not all of the answers are apparent when working on complex projects and should strive to adapt to improve systems and project management processes to navigate and overcome challenges [50].

In system design and abstraction, trade-offs refer to the compromises made when prioritizing certain aspects of a system over others. When a complex system is abstracted, it is simplified to improve understanding and manageability, which often means sacrificing specific details. For example, creating a high-level abstraction, such as a user-friendly interface, in software design may hide complex backend processes, making it easier for users but limiting direct control over certain functionalities. Similarly, simplifying a mechanical model to highlight key forces in engineering can speed up analysis but might overlook minor stresses that could cause issues later. The key idea is that gaining simplicity and clarity often comes at the cost of losing some detail and specificity. Designers must carefully decide what to keep and what can be omitted to ensure the system remains functional and practical. Finding the right balance is essential for successful design and usability [125].

Choosing between abstraction hierarchy and hierarchical abstraction comes down to aligning each method with the project's intended goals and the complexity of the intended use. The abstraction hierarchy excels in structured analysis, offering clear traceability between functional intent and physical implementation, making it ideal for design, diagnostics, and operator understanding. While aiming for greater flexibility in development makes hierarchical abstraction broadly applicable, it requires discipline to maintain clarity and consistency across abstraction layers [126].

Both approaches simplify complexity in different ways, and their effectiveness depends on thoughtful application. Understanding when and how to apply each supports building functional, scalable, adaptable, and intelligently designed building systems, emphasizing that merely using these methods is insufficient. Developers must understand and apply the technique intentionally based on the system's needs and project context to be effective [90].

Abstractions are highly dependent on the subject area and can generally be grouped into two categories: models associated with the subject and models within the topic. The first type focuses on system dynamics, including how processes are measured and how systems respond, whether in real-world applications or simulated environments. The second type deals directly with raw sensor data and the observable behaviors of causal mechanisms within the system itself.

These two forms of abstraction bring forward two fundamental attributes: causality and composability [90].

Causality refers to how relational dependencies between components determine outcomes, while composability captures how system elements are structured and interact to form a coherent whole. Composability, on the other hand, describes how everything within the system is connected, inherently implying a hierarchical structure that reflects how components fit and function together [90].

Causality defines the relationships between cause and effect, serving as a framework for understanding how components within a system interact and influence one another, ultimately driving overall behavior [53]. In systems engineering, recognizing causal dependencies, where the function of one subsystem directly affects another, is essential for identifying critical paths and potential bottlenecks. Understanding these relationships helps clarify functional dependencies, optimize overall system performance, and develop strategies to mitigate risk [127].

Composability complements causality by defining how elements within a system connect and interact. All technologies, including communication networks, electrical grids, and transportation systems, are interconnected, a characteristic of modern technology. A composable system should be built from well-defined components or subsystems that integrate seamlessly to deliver the required functionality and reflect the principles of a modular design. This systems engineering strategy organizes elements into logical, purpose-driven units.

Practical modularity improves system clarity, enabling developers to scale or extend functionality by reusing and replicating proven components. This means that systems designed with composability can expand or evolve without disrupting their overall structure.

For example, new features like a personalized recommendation engine or additional language support can be integrated into a modular streaming platform like Netflix. Each component is designed to align with a consistent system structure; the platform continues to operate reliably as it scales and provides additional features, regardless of user demand.

Adding new features must ensure system expansion does not compromise structural integrity or introduce unnecessary complexity. Designing with composability helps prevent the system from becoming a tangled web of interdependencies. Instead, it becomes a well-organized collection of independent yet interoperable services such as Spotify, Netflix, and a sports channel working together like a finely tuned engine to deliver a unified outcome: entertainment on demand, assuming connectivity is in place [109].

1.12 Finding the Right Balance: The Challenges of Abstraction in System Design

From a systems perspective, abstraction brings its own set of challenges. Over-reliance on abstraction can obscure understanding and introduce unnecessary complexity. On the other hand, insufficient abstraction can lead to tightly coupled systems with limited flexibility and high interdependency. Striking the right balance is crucial for capturing essential details while filtering out nonessential information, maintaining clarity, and adapting to system design [109].

1.12.1 Dependency Management

Effective dependency management starts with identifying the various categories and types of dependencies that influence project execution. The main categories of dependencies are: logical planning dependencies, which determine the required sequence of tasks; resource-based dependencies, which rely on the availability of critical resources; and preference-based dependencies, which are influenced by strategic decisions rather than absolute requirements. Strategic choices are deliberate decisions designed to support long-term objectives, maximize resource efficiency, reduce risks, and enhance competitive advantage [53]. Within these categories, there are four main types of dependencies [128]:

- Types: Finish-to-Start (FS): A task must finish before the next one starts (most common).
- Start-to-Finish (SF): A task cannot finish until another one starts (rare).
- Start-to-Start (SS): Two tasks must begin simultaneously.
- Finish-to-Finish (FF): Two tasks must be completed together.

An example of the four dependencies focused on each dependency type is as follows:

- Finish-to-Start (FS): Example: A database update must be completed before the application can use the new database. The application will rely on a new approach to function correctly, so it cannot start processing queries until the update is done.
- Start-to-Finish (SF): Example: In a data center migration, the backup system cannot be deactivated until the new system is online and fully operational. This ensures continuity of service during the transition from the old system to the latest.
- Start-to-Start (SS): When deploying a distributed application, the authentication and session management services must be launched simultaneously to enable seamless user login and session validation. Users may encounter login failures or session inconsistencies if one starts without the other.
- Finish-to-Finish (FF): Example: In a software release pipeline, automated quality assurance tests and security vulnerability scans must be completed before initiating the production deployment. Both processes must be finished to ensure code integrity and application security in the live environment.

Understanding these distinctions allows project managers and teams to visualize and map out dependencies effectively. Dependency mapping is the foundation of effective dependency management, clarifying task sequencing, resource allocation, and potential bottlenecks. Recognizing these differences informs how developers can visualize, map out, or handle the dependencies. Visualizing dependencies through dependency mapping becomes the first step in practicing dependency management [53].

Dependency management is about organizing and controlling the relationships between system components. Effective management improves modularity, maintainability, and adaptability. Learning from past successes and failures enables developers to make more informed decisions, ultimately leading to improved outcomes. Management is not about rigid structures, but about understanding and strategically managing dependencies for efficiency and resilience [129].

Abstraction and hierarchical system understanding benefit considerably from effective dependency management because they rely on clearly defined relationships between system components [130]. When dependencies are well-managed:

- Abstraction becomes more meaningful: Developers can better abstract lower-level details without losing track of how components interact because dependencies are clearly defined and organized. This enables a systemic approach, making it easier to isolate functionality and work at the right level of detail [129].
- Hierarchical structures remain coherent: A systemic approach views a system as a layered or modular architecture, where components are organized in structured tiers or modules to support clarity, manageability, and scalability. Unmanaged dependencies can make understanding the boundaries between layers difficult, leading to ambiguity and confusion. Identifying dependencies at each layer or module is essential to preserving its function and maintaining a clean, understandable system hierarchy [131].
- Reengineering and scaling are safer: If dependencies are accurately understood, changes to one component are less likely to unintentionally impact others. This directly supports abstraction by enabling modularity, improving clarity, and reinforcing the separation of concerns within the system [132].

Abstraction cuts down complexity, hierarchies add structure, and dependency management keeps relationships clear and intentional. Together, they make systems cleaner, more modular, and easier to understand.

By leveraging dependency management, developers can prevent tight coupling between components, making the system more modular and more straightforward to evolve. Striking the right balance between abstraction and hierarchy ensures that the system remains flexible enough to accommodate future changes while maintaining clarity and organization [133].

Effective dependency management is important in system design, particularly when coordinating interactions between subsystems. When interfaces are poorly defined, problems can arise; for example, if a sensory subsystem changes its data format and the control subsystem isn't prepared for that change, the system may fail. Tight coupling between components, where one relies heavily on another's internal workings, makes updates difficult and costly. A more reliable approach is to establish clear interfaces that define the data exchanged and how communication occurs. This allows each part of the system to evolve independently, as long as the agreed interface remains consistent. The result is a more adaptable, easier-to-maintain system that can handle change without unnecessary disruption [133].

In systems development, abstraction and hierarchy depend heavily on well-defined dependencies. Abstraction simplifies complexity by emphasizing essential features while hiding unnecessary details. Dependencies indicate which components or layers rely on others, enabling developers to understand system structure and implement changes without unintended consequences.

Hierarchy further organizes the system into levels of abstraction, where higher-level functions are built upon lower-level implementations. Explicit dependencies between these levels help maintain structural coherence and design integrity [133].

For instance, effective dependency management is essential to achieving a systems' intended functionality. A clear understanding of relationships between system elements allows teams to make changes with confidence. Managing these interconnections involves analyzing dependencies from both top-down and bottom-up perspectives, which can be considered as key strategies for understanding how different elements interact within the systems' architecture [133].

A dependency occurs when one task or component relies on the output or completion of another. In system development, components designed to perform specific functions often depend on each other to operate as intended. As the number of dependencies increases, so does the complexity involved in managing them. This challenge is amplified in agile development, where tasks and system structures emerge dynamically throughout the project. In this context, dependencies arise frequently, as the progress of one component or subsystem often hinges on the timely delivery of another [133].

As systems become increasingly complex, many interdependent components must evolve simultaneously, making coordination and integration significantly more challenging. Agile's emphasis on flexibility adds another layer of complexity, as the continuous evolution of system structures leads to the constant emergence of new and shifting dependencies throughout the development process [134].

A dependency exists when one task or component must wait for another to be completed before it can proceed. In system development, individual elements are designed to fulfill specific functions, but they often depend on other components to operate as intended. As the number of these dependencies increases, they become more challenging to track and can pose risks to the project's overall success. This challenge is especially evident in agile development, where plans and structures are frequently adjusted. In such a dynamic environment, dependencies can emerge unexpectedly, as teams often rely on deliverables from others to move forward.

The complexity increases further when multiple teams work in parallel, building and integrating different parts of the system simultaneously. Agile methods compound this by introducing continuous shifts in priorities and task assignments, which can lead to new and unforeseen dependencies throughout the development process.

Effectively managing dependencies is not just a technical function but part of delivering functioning systems. As development becomes more distributed and adaptive, the ability to recognize, coordinate, and respond to shifting dependencies becomes essential. Regardless of the approach, whether agile or traditional, what truly matters is effective communication, collaboration, and adaptability when circumstances change. Managing complexity means designing systems that remain functional, stable, and aligned with objectives, even as conditions change [135].

1.13 Managing Complexity Through Layered Structures

Organizing relevant information into a functional structure is essential for managing the complexity of large-scale systems, which often comprise multiple interdependent layers. Each layer builds on the one below it to deliver overall functionality. In software architecture, hierarchical layering is a common strategy for structuring systems and managing complexity. However, as layers become entangled and dependencies spiral out of control, the architecture weakens.

When analyzing complex systems, it is common to find that their structure spans several layers, all linked by various dependencies. Understanding how these dependencies interact requires more than surface-level observation; it calls for deliberate and focused analysis. Since each layer can influence the others, even small changes may lead to cascading effects. Practical dependency analysis reveals bottlenecks, integration issues, and risks threatening system stability. Without it, developers struggle to predict component interactions, raising the risk of failure. These interconnections must be analyzed systematically to optimize performance and support a smoother development and integration process [136].

A clear example of this can be seen in modern system-on-a-chip (SoC) design. For many years, engineers have relied on structural hierarchies to manage the inherent complexity of chip design. While this approach has its limitations, and no universal solution exists for breaking down complex problems, it remains effective when guided by the specific details of the system and the objectives of the analysis [136].

System layering can be viewed through multiple hierarchies, such as functional, architectural, or dependency-based, each offering a valid perspective. Together, they form a comprehensive view of the system. Integrating these perspectives enables clearer analysis, better organization, and improved maintainability and scalability. Consider the following (Table 1.5) [137–139]:

When system components become overly interconnected, layering can become entangled, making the system difficult to manage, modify, or understand. As a result, dependencies can spiral out of control, making stability increasingly complex to maintain. This challenge is especially apparent in integrated circuit (IC) design, where the scale and complexity of modern chips require careful dependency management across multiple

Table 1.5 Hierarchical perspectives of system analysis

Perspective	Focus	Key elements
Functional view [137]	The system's functional elements, their responsibilities, interfaces, and primary interactions	Guide the shape of other system structures, such as information, concurrency, and deployment structures
Architectural view [138]	Includes the logical, development, process, physical, and scenarios views,	Providing a comprehensive framework for understanding system architecture
Dependency view	Analyzing dependencies among architectural views is crucial for understanding the interrelations within a system	Address system dependencies and integration

functional blocks. As chip designs become more complex, algorithms often require longer execution times. A practical approach is to break them into smaller, manageable parts, which not only improves execution speed but also reduces resource usage. By addressing dependencies effectively, designers can optimize performance, improve scalability, and streamline development workflows [136].

Effectively managing this entanglement requires methods to solve a complex puzzle. After all, management involves decision-making [140].

Structural hierarchy helps manage complexity in chip design by organizing components into layers, allowing designers to break down the chip's complexity into manageable segments. Although the informal structural hierarchies commonly used in semiconductor design have limitations, they generally meet the needs of most projects. However, as chip architectures become increasingly advanced, a single hierarchical structure may no longer suffice. In such cases, additional hierarchies will be needed to address challenges that traditional approaches cannot fully resolve. These alternative hierarchies enhance design flexibility, improve performance, and facilitate smoother integration, ensuring that increasingly complex chips remain scalable, efficient, and maintainable [136].

In chip design, information flows through layered structures, with each level performing increasingly complex functions. A bottom-up process enables higher layers to build on signals from lower ones, while top-down feedback provides control and adjustments, guiding the behavior of underlying components. This two-way flow of data and control supports system stability and efficiency [136].

This layered approach is part of what is called structural decomposition, which creates a clear hierarchy to manage complexity and organize system components. However, late-stage modifications can introduce challenges. Instead of reworking the entire structure, engineers often make localized changes to maintain progress. This practice, known as informal structural decomposition, can lead to inconsistencies but remains widespread due to its practicality. Design tools assist in managing these issues and preserving overall system integrity [136].

From a systems engineering perspective, structural decomposition is a way to manage complexity. However, it relies on certain assumptions and design choices that can restrict development later. Decisions made early, such as how module interfaces or data pathways are structured, can complicate the addition of new features or technologies later. When requirements change, these initial choices can result in costly redesigns or temporary workarounds to maintain smooth functionality.

Despite its practical benefits, this process can lead to a highly entangled system, where growing interconnections between components make management and coordination significantly more difficult. As systems become more tightly integrated, the task of understanding, evaluating, and maintaining these dependencies becomes increasingly complex and resource-intensive.

Tightly coupled systems complicate fault isolation, slow integration, and weaken abstraction and modularity. From a systems engineering perspective, poor understanding or unmanaged dependencies can compromise modularity, reduce system flexibility, and introduce risks during integration, testing, and long-term maintenance. In many projects, managing dependencies becomes a critical task within the overall development process. This requires a focused analysis process and well-defined strategies to maintain system development integrity and minimize risks [136].

In systems engineering, impact analysis is crucial for evaluating the potential effects of proposed changes within a system. It involves systematically identifying, analyzing, estimating, and addressing the impact of significant technical changes that may ripple across various components. Effective impact analysis helps anticipate risks, understand dependencies, and plan necessary adjustments to maintain system stability and integrity [141].

This process is central to effective change management, since it supports the systematic evaluation of how modifications to one component may impact other system elements, helping to prevent adverse effects on functionality, performance, or integration. By employing impact analysis, systems engineers can maintain system integrity, manage risks, and ensure that the system meets its intended requirements and performance parameters [142].

Managing dependencies in complex, layered architectures requires a balance of modularity, autonomy, and communication. The Dependency Rule, a core principle of Clean Architecture, states that dependencies should always point inward toward higher-level policies and abstract interfaces, not concrete implementations. Ensuring that inner layers representing domain logic and business rules remain independent of outer layers, such as infrastructure and technical details, promotes flexibility and adaptability to change. This inversion of dependencies enforces decoupling between layers, making the system more maintainable and responsive to changes [142].

By selecting suitable design patterns, implementing thorough testing strategies, and cultivating a culture of continuous improvement, development teams can navigate the complexities of modern software and hardware development, ultimately building functional, maintainable, and scalable systems over time. After all, managers are constantly faced with complex challenges, making it essential to develop a structured understanding based on what can be analyzed and controlled [136].

Effective dependency management supports the development process by providing an understanding that isn't about knowing how system parts interact. Abstraction helps manage complexity by exposing only what is necessary. Hierarchy organizes those abstractions so that the system maintains the necessary functionality. Modular, layered systems enable changes to one part without disrupting the rest, thereby improving long-term planning and iteration. In systems engineering, modular, layered architectures would allow changes to individual components without disrupting the entire system, supporting better long-term planning and iterative development [143].

1.14 From Problem Definition to Solution Design

The problem space encompasses all aspects of a problem, including its context, stakeholders, requirements, and constraints. It represents the current state and the desired future state. In contrast, the solution space examines all potential solutions, ranging from minor updates to significant capability innovations. This phase involves designers evaluating various approaches to address the challenges defined in the problem space [144].

A well-defined problem space is essential for effectively navigating the solution space. Within the design space, both viable and non-viable designs exist, but optimization techniques focus on identifying feasible solutions. The solution space, a subset of the design space, can be explored using optimization algorithms, such as box and polytope optimization, as well as other methods, to locate optimal design configurations. Design patterns help bridge the problem and solution spaces by providing templates that capture design knowledge and proven solutions to recurring challenges, making it easier to apply best practices across various scenarios [145].

Understanding the relationship between problem and solution spaces is crucial in systems engineering. Establishing this connection helps identify practical and workable solutions while keeping design efforts aligned with real-world needs. Maintaining that link ensures design processes remain flexible and responsive to changing requirements and constraints.

When starting a new project, systems engineers often employ hierarchical design approaches to manage the complexities of both the physical and digital spaces. This approach breaks systems into smaller, more manageable parts, making it easier to understand how system components interact. With a clearer understanding of these interactions, developers can design robust and adaptable solutions that better meet the evolving needs of stakeholders [28, 145].

Understanding the relationship between the problem and solution space increases the potential for developing a realizable solution, establishing insight into practical, implementable solutions within the solution space [145]. The solution space enables the exploration of product performance outcomes and the dynamic factors that influence concurrent development processes. This can be used to explain conditions under which concurrency is a desirable strategy [146].

Achieving a robust solution involves identifying its position within the feasible operating regions and assessing its sensitivity to changes in controllable variables [147]. The design process provides a cognitive framework of analysis, synthesis, and evaluation, which many professional designers use to solve design problems. Developers discover through testing, revising, and retesting that while some solutions may work better than others, there are many possibilities for success.

The concepts of problem space, solution space, open system, and closed system are central to understanding the engineering process. These concepts should be explained using systems theory as the foundational theoretical framework. In-depth examples support this articulation by covering various needs and requirements, and providing step-by-step insights into how elicitation, derivation, and decomposition occur during problem formulation [144].

1.14.1 The Abstraction in Systems Engineering

Abstraction is a key concept in computer science that enables software developers to manage complexity [148]. Developers can apply solutions from one problem to others, making it easier to understand the system's structure and how its parts connect. This makes it simpler to integrate components, spot bottlenecks, and focus on what matters without being distracted by unnecessary details. Abstraction also supports modular design, simplifies troubleshooting, and makes projects easier to scale.

Developing complex systems requires a deep understanding of both problem and solution spaces. Computational thinking skills, such as abstraction, pattern recognition, and modeling, help engineers understand complex systems and solve problems efficiently. Hierarchical dependencies, causality, and composability show how different parts are connected, highlighting the need for apparent abstraction [149].

Abstraction simplifies complex problems, identifies common patterns, and enables solutions that can be applied in various situations. This broader perspective makes it easier to understand how system components work together and supports the creation of stronger, more flexible designs.

In optimization, solution spaces can be explored using box and polytope optimization algorithms, which focus on maximizing the range of workable designs. Box optimization operates within defined upper and lower bounds, researching potential solutions, while polytope optimization navigates through a geometric space shaped by linear constraints, optimizing along its vertices and edges. These methods are valuable when applied to design abstractions in human–computer interaction, as described, where varying levels of abstraction in system design, such as e-mail interfaces, benefit from optimization for efficiency and user experience. Design patterns correlate problems and solutions, deriving practical design concepts and providing templates for emerging challenges. Reusing these patterns accelerates development and promotes consistency across projects. By applying box and polytope optimization methods, design patterns can be fine-tuned for efficiency, scalability, and structured solutions that adhere to abstract design principles in human–computer interaction [150].

From a cognitive perspective, systems thinking is important for user-centered design. It allows engineers to predict the effects of new human–computer interactions. Abstraction helps by focusing on the essential aspects of a problem while disregarding unnecessary details, resulting in more straightforward and more efficient designs [150].

Understanding the relationship between problem and solution spaces is essential for effective systems engineering. Practical, implementable solutions become achievable while keeping designs adaptable to changing requirements, dependencies, and constraints. Hierarchical design systems build on this by reducing complexity and clarifying the interactions between components. This structured approach results in solutions that are resilient, flexible, and capable of meeting stakeholders' evolving needs [144].

The concept of problem and solution spaces is fundamental in beginning systems development. It provides a structured approach to addressing design and development challenges. The problem space focuses on understanding stakeholders' needs, goals, dependencies, and constraints, establishing a clear understanding of the proposed development [151].

This involves recognizing the specific requirements and examining the context, assumptions, and conflicts that may influence how the problem is defined [151, 152]. Conversely, the solution space includes a range of potential design alternatives that can meet the established requirements, supporting the selection and configuration of specific options based on predefined specifications [151].

The system engineer defines the solution, so an ill-defined problem results in a meaningless solution. Continuous refinement and adaptation enable ongoing adjustments as new information becomes available, ensuring the final solution remains aligned with the evolving understanding of the problem. An early focus on defining the problem, managing information, and understanding the design's systematic nature is crucial, as misunderstandings can impact solution effectiveness due to cognitive (thinking, understanding, learning, and remembering) and metacognitive (awareness and regulation of one's thinking) factors.

Design optimization is more effective when the solution space, or the range of all possible designs, includes only practical and feasible options. This arrangement accelerates development and enables various design elements to be improved independently [153].

Methodically constraining the solution space accelerates optimization and promotes independent progress across different areas of system design [154]. This targeted approach stands in contrast to traditional methods, which often struggle with broad, undefined solution spaces, leading to extended development timelines and suboptimal outcomes [155]. Two key methodologies used to operate within this refined design space are box optimization and polytope optimization, each suited to specific types of constraint relationships.

Box optimization treats the solution space as a hyper-rectangle, where each design variable is bounded independently by upper and lower limits. This method is most effective when constraints are unrelated and can be addressed separately. In contrast, polytope optimization defines the solution space through a set of linear inequalities, allowing for more flexible navigation of interdependent constraints. This approach excels when design variables are closely linked and must be evaluated collectively [155].

Together, these methods enable engineers to maximize the feasible design space, explore a broader range of options, and ensure compliance with all system requirements. They also support parallel development by establishing well-defined boundaries, allowing subsystems to evolve independently while maintaining overall coherence.

Design patterns complement these methods by bridging the gap between the problem and solution spaces. They capture proven design strategies in reusable templates, providing consistent and effective responses to common engineering challenges. Recognizing this connection helps developers generate solutions that are not only technically sound but also practical and implementable within the defined solution space [155].

1.15 Structured Abstraction for System-Level Evaluation

In systems engineering, evaluation frameworks must address not only technical performance but also guide developers throughout the system lifecycle. As modern systems become increasingly complex and incorporate more automation and advanced human–machine interfaces, traditional performance-based evaluations may be insufficient for managing their development. To support the analysis required to evolve a system from concept to realization, incorporating a model that combines hierarchical structuring with abstraction modeling could help frame both system performance and the developer experience, enabling a deeper and more comprehensive understanding through a structured evaluation framework [156, 157].

A model would build on traditional evaluation, which emphasizes performance and effectiveness, but extends further to include pragmatic evaluation, an approach that examines how a system operates in real-world contexts. Pragmatic evaluation focuses on usability, user interaction, and, when applicable, the cognitive impact of automation or autonomy. Both perspectives align with core principles of human-systems integration and underscore the importance of evaluating not only technical system performance but also user-centered factors [158].

To support the development of the model, two key types of metrics are used: those that evaluate objective system performance and those that capture user experience and real-world interaction with the system. This dual focus ensures a balanced assessment, addressing both technical functionality and user-centered outcomes. The metrics are outlined as follows:

(a) Traditional metrics focus on measurable system outputs such as performance, effectiveness, and reliability. Traditional metrics emphasize quantifiable outcomes such as task completion, accuracy, and response times, which are fundamental to early-stage system validation and verification activities in systems engineering [159].
(b) Pragmatic metrics assess how the system performs in practical use, encompassing usability, user interaction, and the cognitive demands placed on users in operational contexts. Pragmatic metrics expand the scope to include usability, user interaction, and cognitive impact dimensions, which are tied to how the system is perceived and experienced by human operators [160, 161]. These measures are important for capturing real-world use cases and understanding the user cognitive requirements during system interaction in a required operational setting. A two-tiered evaluation approach supports standard systems engineering practices by addressing both the functional capabilities of the system and its performance within operational contexts [87].

At a higher level of abstraction, evaluation focuses on subjective user perceptions, including trust, understanding, assessment of efficiency, and satisfaction. These indicators help assess how well users internalize and work with system behavior. Trust, for example, plays a central role in system adoption and appropriate reliance, particularly in complex systems that incorporate automated decision-making or intelligent control components [162]. Understanding supports the formation of effective mental models, which in turn affect both task efficiency and error mitigation [163]. Satisfaction measures perceived value and overall experience, which are key to long-term system use.

At this level, the impact of technical and interface design decisions on user cognition and system usability becomes evident, particularly in supervisory control, decision-support tools, and safety–critical systems [164].

At the most abstract level, the model assesses the system's ability to justify its outputs through recommendations, explanations, and uncertainty quantification. These outputs support the need for traceable system behavior, fostering operator confidence and informed decision-making in complex operational environments [165, 166].

The engineering development process must ensure that the system's decision-making logic aligns with design objectives and satisfies defined development requirements. Systems in which the decision-making process is clear to the developer are more likely to yield reliable and predictable behavior. This clarity fosters the development of calibrated trust and helps maintain situational awareness, especially in safety–critical operational environments [167]. Additionally, representing uncertainty in system outputs enables risk-informed decision-making, which is integral in fields such as aerospace, medicine, and defense [168]. Finally, the model incorporates self-assessment of decision logic, allowing the developer to understand how the system needs to perform and how it evaluates information to arrive at the rationale for its performance or decision-making.

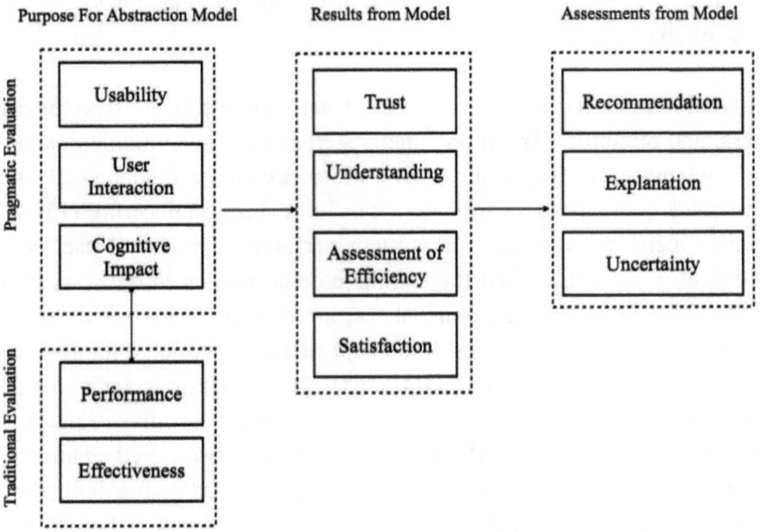

Fig. 1.3 Hierarchical abstraction in systems evaluation [90]

To help solidify the developers' understanding of how system behavior should be assessed from both technical and user-centered perspectives, an evaluation framework must be established across three levels of abstraction. The first level focuses on objective outputs at the system level, referring to quantifiable performance measures such as task accuracy, response time, and functional reliability. The second level addresses subjective experience at the user level, emphasizing human perceptions such as trust, satisfaction, and usability, which influence how users may interact with and interpret the system during operational use. The third level includes epistemic self-assessment of decision logic, where the system evaluates and communicates its reasoning and confidence in its outputs.

Epistemic refers to things related to what a system knows, does not know, or believes it knows, making this form of self-assessment important for identifying system limitations and supporting informed decision-making. This layered structure facilitates a more comprehensive understanding of system behavior by integrating technical performance, user experience, and system-level evaluation.

Each level builds on the one before it, following systems thinking principles that emphasize continuous feedback and alignment between how a system works and what it needs to achieve over its lifecycle [87, 169]. To be successful, system development must incorporate analytical processes that systematically support design choices that are traceable, testable, and aligned with both technical requirements and operational needs.

This framework supports iterative development for systems engineers by integrating human feedback, usability validation, and model transparency. Such an approach aligns with systems engineering lifecycle models that prioritize stakeholder engagement, verification, and adaptability [86]. By addressing factors such as developer understanding, interface reliability, and clarity of system rationale, this model is especially relevant for autonomous systems, cyber-physical systems, and intelligent decision-support tools (Fig. 1.3).

References[1]

1. McLeod, S. (2024). https://www.sciencedirect.com/science/article/abs/pii/B9780444506498500127?via=ihub%20learning%20styles%20and%20experiential%20learning%20cycle; https://www.simplypsychology.org/learning-kolb.html
2. Krueger, K. L., Diabes, M. A., & Weingart, L. R. (2023). *The psychological experience of intragroup conflict*. Tepper School of Business, Carnegie Mellon University. https://safe.menlosecurity.com/doc/
3. Kamarudin, K. M., Ridgway, K., & Ismail, N. (2016). Abstraction and generalization in conceptual design process: Involving safety principles in TRIZ-SDA environment. *Procedia CIRP, 39*, 16–21.
4. Johnson, S. (2010). The genius of the tinkerer, September 25, 2010 WSJ
5. Hardwick, T. (2023). *Business and enterprise architecture & strategy.* https://quantumfields.co.uk/architecture/harnessing-complexity-the-power-of-systems-thinking-in-engineering#:~:text=As%20engineering%20projects%20become%20increasingly,drive%20outcomes%20with%20unforeseen%20consequences
6. Six Sigma. (2024). Applications of systems thinking—Systems thinking framework. https://www.6sigma.us/systems-thinking/systems-thinking-framework/
7. Sulistyawati, K. (2023). *Systems mapping tools & design levers.* https://medium.com/somiacx/systems-mapping-tools-design-levers-e9f8b076ffd9
8. Sevaldson, B. (2017). *Leverage point investigation—Systems oriented design.* https://systemsorienteddesign.net/leverage-points/
9. Kim, B., Cruden, G., Crable, E. L., Quanbeck, A., Mittman, B. S., & Wagner, A. D. (2023). A structured approach to applying systems analysis methods for examining implementation mechanisms. *Implementation Science Communications, 4*(1), 127. https://doi.org/10.1186/s43058-023-00504-5
10. Systems Thinking. (2023). *How feedback loops can cause system failures.* LinkedIn. https://www.linkedin.com/advice/3/what-some-examples-feedback-loops-have-led-unintended
11. Ionescu, D. (2022). *What is induced demand?* | Planopedia—Planetizen. https://www.planetizen.com/definition/induced-demand
12. How algorithms are amplifying misinformation and driving a wedge ... (2021). https://thehill.com/changing-america/opinion/581002-how-algorithms-are-amplifying-misinformation-and-driving-a-wedge/
13. Saplagio, A. (2024). *The role of algorithms in amplifying misinformation.* Comms Room. https://commsroom.co/the-role-of-algorithms-in-amplifying-misinformation/
14. Its' the algorithm: A large-scale comparative field study of ... (2023). https://knightcolumbia.org/content/its-the-algorithm-a-large-scale-comparative-field-study-of-misinformation-interventions
15. Cho, C. Y., & Harris, L. (2023). Social media algorithms: Content recommendation, moderation, and congressional considerations (CRS product). H.R. 2154; S. 5339. Library of Congress. https://www.congress.gov/crs-product/IF12462

[1] Abstractions and Hierarchical

16. Meadows, D. H., & Wright, D. (2008). *Thinking in systems: A primer* (p. 146). Chelsea Green Publishing.
17. Analytical Reasoning. (2023). *Leverage points: How to influence complex systems.* LinkedIn. https://www.linkedin.com/advice/0/how-do-you-identify-leverage-points-interventions
18. Ehrlichman, D. (2018). *Identifying leverage points in a system.* https://medium.com/converge-perspectives/identifying-leverage-points-in-a-system-3b917f70ab13
19. Zahaib Nabeel, M. (2024). AI-enhanced project management systems for optimizing resource allocation and risk mitigation: Leveraging big data analysis to predict project outcomes and improve decision-making processes in complex projects. *Asian Journal of Multidisciplinary Research & Review, 5*(5), 53–91. https://doi.org/10.55662/AJMRR.2024.5502
20. Zalando Product Design. (2024). *The power of metaphor in designing a visually elevated experience.* Zalando Design. https://medium.com/zalando-design/the-power-of-metaphor-in-designing-a-visually-elevated-experience-6311f4596ff.
21. Lai, M. (2023, October). Experience designs' transformation towards experience-driven transformation: A practical perspective. In *IASDR 2023: Life-changing design.* https://www.researchgate.net/publication/374552590_Experience_design's_transformation_towards_experience-driven_transformation_a_practical_perspective
22. Tekir, G. (2023). *XP Metaphor*. Project Management Knowledge. https://medium.com/project-management-knowledge/xp-metaphor-6cf2e0368328
23. Whalen, M. W., Gacek, A., Cofer, D. D., Murugesan, A., Mats, P. E., & Rayadurgam, S. (2013). Your "what" is my "how": Iteration and hierarchy in system design. *IEEE Software, 30*(2), 54–60. https://www.researchgate.net/publication/259932771_Your_What_Is_My_How_Iteration_and_Hierarchy_in_System_Design
24. Allen, R. (2018). *Using hierarchy in complex system requirements and design.* https://accendoreliability.com/using-hierarchy-in-complex-system-requirements-and-design/
25. DocSheets. (2024). *Requirements management: A comprehensive guide.* https://www.docsheets.com/requirements-management-guide/
26. Shevchenko, N. (2021). *Requirements in model-based systems engineering (MBSE).* https://insights.sei.cmu.edu/blog/requirements-in-model-based-systems-engineering-mbse/
27. Pohl, K., & Sikora, E. (2007). *The co-development of system requirements and functional architecture.* https://www.researchgate.net/publication/278714556_The_Co-Development_of_System_Requirements_and_Functional_Architecture
28. Chapter 2: Systems engineering (SE)—The systems design process. (2023). https://www.eng.auburn.edu/~dbeale/ESMDCourse/Chapter2.htm
29. Systems Engineering. (2023). *Scaling and simplifying models for systems engineering.* LinkedIn. https://www.linkedin.com/advice/0/how-do-you-scale-simplify-your-models-systems
30. Buede, D. M., & Miller, W. D. (2016). *The engineering design of systems: Models and methods* (pp. 12, 20, 38, 54). Wiley. http://alvarestech.com/temp/capp/IDEF0/Idef0-SYSML.pdf
31. Federal Aviation Administration (FAA). (2024). *Aircraft systems* (Chap. 7). https://www.faa.gov/sites/faa.gov/files/09_phak_ch7.pdf
32. Chapter 2: Systems engineering (SE)—The systems design process. (2018). https://www.eng.auburn.edu/~dbeale/ESMDCourse/Chapter2.htm
33. Systems Engineering—Intelligent Motion Lab. (2016). Section VII. *Robotic systems in practice.* https://motion.cs.illinois.edu/RoboticSystems/SystemsEngineering.html#Section-VII.-ROBOTIC-SYSTEMS-IN-PRACTICE
34. Caminao's Ways. (2017). *Abstractions-based systems engineering.* https://caminao.blog/system-engineering/models-perspectives/abse/

35. GeeksforGeeks. (2023). *Why are abstractions important in system design?* https://www.geeksforgeeks.org/why-are-abstractions-important-in-system-design/
36. Puik, E. (2024). *Abstraction in engineering: Emphasize the key points of your system.* https://www.linkedin.com/pulse/systems-engineering-tidbits-abstraction-emphasise-key-erik-puik-78nue
37. Wheatcraft, L. (2017). *Levels of abstraction when writing requirements.* ArgonDigital. https://argondigital.com/blog/product-management/levels-of-abstraction/
38. University of Oregon. (2023). Procedural and declarative knowledge. In *Approaches to (language) learning* (Chap. 2). https://opentext.uoregon.edu/languagelearningedition1/chapter/procedural-and-declarative-knowledge/#:~:text=As%20shown%20in%20the%20illustration,109
39. SystemsEngineering—Intelligent Motion Lab. (2016). *Systems engineering.* https://motion.cs.illinois.edu/RoboticSystems/SystemsEngineering.html
40. Olmstead, L. (2022). *7 types of organizational structures + examples.* Whatfix. https://whatfix.com/blog/organizational-structure/
41. Eby, K. (2021). *Essential guide to project organizational structure.* Smartsheet. https://www.smartsheet.com/content/project-management-organization?srsltid=AfmBOornCIPzX9GiwOo3P-4mfXD-JcE2ZVpettEISM97y5lYBm8t6Dh
42. Mast, J. (2024). *7 tips to minimize ambiguity in prompts.* White Beard Strategies. https://whitebeardstrategies.com/blog/7-tips-to-minimize-ambiguity-in-prompts/
43. HabileData. (2024). *How to remove bias and ambiguity in text annotation for machine learning.* Medium. https://habiledata.medium.com/how-to-remove-bias-and-ambiguity-in-text-annotation-for-machine-learning-95e21c108ff8
44. Lcom Team. (2025). *Understanding abstraction: Everyday examples.* Learning.com. https://www.learning.com/blog/examples-of-abstraction-in-everyday-life/
45. CelerData. (2024). *A guide to data abstraction and its significant benefits.* https://celerdata.com/glossary/a-guide-to-data-abstraction
46. Xu, W. (2007). Identifying problems and generating recommendations for enhancing complex systems: Applying the abstraction hierarchy framework as an analytical tool. *Human Factors, 49*(6), 975–994. https://www.researchgate.net/publication/5773247_Identifying_Problems_and_Generating_Recommendations_for_Enhancing_Complex_Systems_Flight_Deck_Automation_Applying_the_Abstraction_Hierarchy_Framework_as_an_Analytical_Tool
47. Vyborov, E. (2022). *The first step to making a decision—Understand the context (Cynefin Framework).* LinkedIn. https://www.linkedin.com/pulse/first-step-making-decision-understand-context-cynefin-vyborov
48. Untools. (2024). *Cynefin framework.* https://untools.co/cynefin-framework/#:~:text=When things aren't under,domain. Act–sense–respond.&text=when cause and effect,domain. Act–sense–respond.&text=to establish some stability,domain. Act–sense–respond.&text=and work on bringing,domain. Act–sense–respond
49. Hinssen, P. (2023). *Dealing with complexity in the never normal.* https://www.peterhinssen.com/blog/dealing-with-complexity-in-the-never-normal
50. Hochstein, L. (2022, September 17). *Up and down the abstraction hierarchy—Surfing complexity.* Retrieved May 8, 2025, from https://surfingcomplexity.blog/2022/09/17/up-and-down-the-abstraction-hierarchy/
51. Audet, L., McEwan, E., & Sargent, E. (2002). *Project interdependency management.* Paper presented at the Project Management Institute Annual Seminars & Symposium, San Antonio, TX. Project Management Institute, Newtown Square, PA. https://www.pmi.org/learning/library/project-interdependency-management-1038

52. Bhuyan, A. (2024). *Effective strategies for software engineers*. Medium. https://aditya-sunjava.medium.com/navigating-dependency-complexity-effective-strategies-for-software-engineers-3835c9d487b3
53. Planview. (2022). *Agile teams: Dependency management and visualization*. https://www.planview.com/resources/guide/what-is-agile-program-management/agile-teams-dependency-management-visualization/
54. EW Solutions. (2025). *The power of abstraction in data modeling*. https://www.ewsolutions.com/power-abstraction-data-modeling/
55. Blanchard, B. S., & Blyler, J. E. (2016). *System engineering management* (pp. 28–32). John Wiley & Sons, Inc.
56. SEBoK. (2024, November 25). Abstraction (Glossary). In *SEBoK v. 2.11*. https://sebokwiki.org/wiki/Abstraction_(glossary)
57. Daga, A., de Cesare, S., & Lycett, M. (2006). *Separation of concerns: Techniques, issues, and implications*. Department of Information Systems and Computing, Brunei University, Oxbridge, Middlesex UB8 3PH, United Kingdom. https://www.researchgate.net/publication/276042443_Separation_of_Concerns_Techniques_Issues_and_Implications/
58. Sulistyawati, K. (2023). *Systems mapping tools & design levers*. Medium. https://medium.com/somiacx/systems-mapping-tools-design-levers-e9f8b076ffd9
59. Fuertes, R. A. (2025). *Agent-based modeling definition: Understanding the basics*. Smythos. https://smythos.com/ai-agents/ai-tutorials/agent-based-modeling-definition/
60. Advancedor Academy. (2024). *Agent-based modeling: Simulating complex systems and interactions*. Medium. https://advancedoracademy.medium.com/agent-based-modeling-simulating-complex-systems-and-interactions-19b71b00bd31
61. Campajola, C., Gangi, D. D., Lillo, F., et al. (2022). Modelling time-varying interactions in complex systems: The score driven kinetic Ising model. *Scientific Reports, 12*, 19339. https://doi.org/10.1038/s41598-022-23770-0
62. Bonabeau, E. (2002). Agent-based modeling: Methods and techniques for simulating human systems. *Proceedings of the National Academy of Sciences of the United States of America, 99*(Suppl 3), 7280–7287. https://doi.org/10.1073/pnas.082080899
63. Kaddouri, A. (2025). *Agent-based modeling techniques: A guide to simulating complex systems*. Smythos. https://smythos.com/ai-agents/agent-architectures/agent-based-modeling-techniques/
64. Naceur, A. (2023). *The role of chaos theory in predictive analytics**. Retrieved from https://amirnaceur.medium.com/the-role-of-chaos-theory-in-predictive-analytics-unraveling-complex-patterns-in-big-data-c0a7a8e0402c
65. GeeksforGeeks. (2024). *Applications of chaos theory*. https://www.geeksforgeeks.org/applications-of-chaos-theory/
66. Bullis, J. (2024). *Scenario planning: Strategies, techniques, & examples*. Cube Software. https://www.cubesoftware.com/blog/scenario-planning
67. Pourrahimian, E., Salhab, D., Shehab, L., Hamzeh, F., & AbouRizk, S. (2024). *Application of chaos theory in project management*. Unpublished manuscript. Retrieved from local file path [file:///C:/Users/fmx06/Downloads/pourrahimian-et-al-2024-application-of-chaos-theory-in-project-management.pdf]
68. O'Donoghue, J. (2023). *A simple guide to hierarchical task analysis*. Make: Iterate. https://makeiterate.com/a-simple-guide-to-hierarchical-task-analysis/

69. Smith, G. D. (2015, October). *The who, what, when, why, and how of agile requirements elicitation.* Paper presented at the PMI® Global Congress 2015—North America, Orlando, FL. Newtown Square, PA: Project Management Institute. https://www.pmi.org/learning/library/who-what-when-why-how-agile-requirements-987; Application of chaos theory in project management. (2024). ASCE Library. https://www.researchgate.net/publication/379046954_Application_of_Chaos_Theory_in_Project_Management
70. Gregor, S., Müller, O., & Seidel, S. (2013, January). Reflection, abstraction, and theorizing in design and development research. In *Proceedings of the European conference on information systems, Utrecht, Netherlands.* https://www.researchgate.net/publication/259639891_Reflection_Abstraction_and_Theorizing_in_Design_and_Development_Research
71. Patterson-Hine, A., Biswas, G., Aaseng, G., Narasimhan, S., & Pattipati, K. R. (2005). *A review of diagnostic techniques for ISHM applications.* https://ntrs.nasa.gov/search.jsp?R=20060019235
72. Sehl, C. (2024, November 22). *Simplicity is key to understanding and achieving goals: Our minds favor simple explanations and efficient actions.* Science Daily. https://www.sciencedaily.com/releases/2024/11/241122130353.htm
73. Agarwal, R. (2025). *Crack ML system design interviews like a pro—Part 1.* https://www.mlwhiz.com/p/crack-ml-system-design-interviews
74. Adu, L. (2023). *Abstraction in software design.* LinkedIn. https://www.linkedin.com/pulse/abstraction-software-design-lawrence-adu
75. Hogan. (2024). *Application of abstraction in system design.* Hogan. B Lab. https://hogantechs.com/en/system-design-abstraction-rpc-grpc/
76. Yoo, D. J. (2018). *Levels of abstraction: A key concept in systems design.* Medium. https://medium.com/@danieljyoo/levels-of-abstraction-a-key-concept-in-systems-design-7fdb33d288af
77. Kalayci, T. (2023). *Abstraction in software: Exploring real-world examples.* Medium. https://medium.com/@tayfunkalayci/abstraction-in-software-exploring-real-world-examples-b2e95d496bef
78. Eshleman, C. (2023). *Multiple layers of abstraction in design systems.* Spotify Engineering. https://engineering.atspotify.com/2023/05/multiple-layers-of-abstraction-in-design-systems/
79. Hashem, Y. (2024). *Abstraction in software: Balancing simplicity and complexity.* Medium. https://medium.com/@yasin162001/abstraction-in-software-balancing-simplicity-and-complexity-abfcefedd1c5
80. Gorin, M. (2024). *Layers and boundaries: The backbone of software architecture.* Medium. https://maxim-gorin.medium.com/layers-and-boundaries-the-backbone-of-software-architecture-e83c8a9342b
81. Miller, D., & Oliver, M. (2015). *Engaging stakeholders for project success.* PMI White Paper. https://www.pmi.org/learning/library/engaging-stakeholders-project-success-11199
82. ProjectManager. (2024). *Work breakdown structure (WBS) guide.* ProjectManager. Retrieved from https://www.projectmanager.com/guides/work-breakdown-structure
83. Systems Engineering Process. (2024). DAU. https://www.dau.edu/acquipedia-article/systems-engineering-process#:~:text=IEEE15288.1%2D2014%20Application%20of%20SE%20on%20Defense%20Programs
84. Shevchenko, N. (2020). *An introduction to model-based systems engineering (MBSE).* SEI Insights Blog. https://insights.sei.cmu.edu/blog/introduction-model-based-systems-engineering-mbse/
85. Blanchard, B. S., & Blyler, J. E. (2016). *System engineering management* (5th ed., pp. 18). Wiley.
86. Blanchard, B. S., & Fabrycky, W. J. (2011). *Systems engineering and analysis* (5th ed., pp. 19–22). Pearson.

87. INCOSE. (2015). INCOSE systems engineering handbook: A guide for system life cycle processes and activities (4th ed., p. 42). Wiley.
88. Blanchard, B. S., & Blyler, J. E. (2016). *System engeering management* (5th ed., pp. 296–305). Wiley.
89. Teachers Institute. (2015). A distinction with a difference: Declarative and procedural knowledge. In *Learning, learner, and development*. https://teachers.institute/learning-learner-development/declarative-vs-procedural-knowledge/.
90. Bencomo, N., Cabot, J., Chechik, M., Cheng, B. H. C., Combemale, B., Wąsowski, A., & Zschaler, S. (2024). *Abstraction engineering*. Cornell University. https://arxiv.org/pdf/2408.14074
91. Dori, D. (2003). Conceptual modeling and system architecting. *Communications of the ACM, 46*(10), 62–65.
92. Wagner, S., & Deissenboeck, F. (2017). *Abstractness, specificity, and complexity in software design*. https://arxiv.org/abs/1709.01304
93. Wigmore, I. (2024). *What is abstraction?* Informa TechTarget. https://www.techtarget.com/whatis/definition/abstraction?utm_source=chatgpt.com
94. Difference between bottom-up model and top-down model. (2022, December 2). GeeksforGeeks. https://www.geeksforgeeks.org/difference-between-bottom-up-model-and-top-down-model/
95. Crudu, A., & MoldStud Research Team. (2024, March 24). *Tips for successful integration of AI-driven personal assistants in software*. https://moldstud.com/articles/p-tips-for-successful-integration-of-ai-driven-personal-assistants-in-software?utm_source=chatgpt.com
96. Effective modular design in software engineering. (2019, December 20). GeeksforGeeks. https://www.geeksforgeeks.org/effective-modular-design-in-software-engineering/?utm_source=chatgpt.com
97. Systems Engineering. (2023). *How to identify and manage dependencies in systems engineering*. https://www.linkedin.com/advice/1/how-can-you-identify-manage-dependencies-systems
98. Cassell, D. (2023). *How these engineers include scalability in every step of development*. https://www.builtinboston.com/articles/-How-Top-Engineers-Approach-Development-With-Scalability-in-Mind
99. Verification and validation: Overview. (2024). AcqNotes. https://acqnotes.com/acqnote/careerfields/verification-validation
100. GeeksforGeeks. (2024). *Introduction to modularity and interfaces in system design*. https://www.geeksforgeeks.org/58ntroduction-to-modularity-and-interfaces-in-system-design/
101. Jovanović, M. (2022). *Clean architecture and the benefits of structured software design*. https://www.milanjovanovic.tech/blog/clean-architecture-and-the-benefits-of-structured-software-design
102. Luthra, A. (2024). *Understanding the dependency inversion principle: Building flexible and maintainable software*. Medium. https://medium.com/@er-abhishek-luthra/understanding-the-dependency-inversion-principle-building-flexible-and-maintainable-software-fa32d117ff4c
103. Coursera Staff. (2024). *What are abstraction layers?* Coursera. https://www.coursera.org/articles/abstraction-layers
104. Mohanan, R. (2024). *Dependency injection: Benefits, types, and real-world examples*. Spiceworks. https://www.spiceworks.com/tech/devops/articles/what-is-dependency-injection/
105. Brebeanu, R. (2022). *What is procedural knowledge and how you can make the most of it?* https://elium.com/blog/what-is-procedural-knowledge-and-how-you-can-make-the-most-of-it/
106. Chi, M. T.H., & Ohlsson, S. (2005). *Complex declarative learning—Arizona State University*. https://www.public.asu.edu/~mtchi/papers/Handbook_Chapter_with_Din.pdf

107. Sumner, S. (2015, April 16). The three systems engineering approaches: Top-down, middle-out, and bottom-up. In *CORE, interfaces, software and solutions, systems engineering.* https://systems-wise.com/the-three-systems-engineering-approaches-top-down-middle-out-and-bottom-up/
108. GeeksforGeeks. (2024). *Dependencies in project management and its types with examples.* https://www.geeksforgeeks.org/dependencies-in-project-management-and-its-types-with-examples/
109. Laoyan, S. (2024). *Understanding dependencies in project management.* Asana. https://asana.com/resources/project-dependencies
110. Datatab. (2024). *Causality: Establishing cause-and-effect relationships.* https://datatab.net/tutorial/causality
111. Freepik. (2024). *Modularity [Icon].* https://www.freepik.com/icon/modularity_17240832
112. The Noun Project. (2024). *Dependency [Icon].* https://thenounproject.com/browse/icons/term/dependency/
113. Dreamstime. (2024). *Scalability [Illustration].* https://www.dreamstime.com/illustration/scalability-icon.html
114. iStock. (2024). *Adaptability [Illustrations].* https://www.istockphoto.com/illustrations/adaptability?page=2
115. ioMosaic. (2024). *Asset integrity management.* https://www.iomosaic.com/services/process-safety-management/asset-integrity-management
116. Medium. (2024). *Structural causal models: A quick introduction.* https://medium.com/causality-in-data-science/structural-causal-models-a-quick-introduction-1ab49259e921
117. Science X staff. (2025). *Simplicity is key: Study finds our minds favor simple explanations and efficient actions.* https://www.msn.com/en-us/news/technology/simplicity-is-key-study-finds-our-minds-favor-simple-explanations-and-efficient-actions/ar-AA1uzTYv?ocid=msedgntp&pc=DCTS&cvid=bcba81469bba449eee76862d8c588be8&ei=37
118. Lin, Y., Zhang, W. J., & Watson, L. G. (2001). On proposal of function-behavior-state framework as refinement of EID framework of human-machine interface design. In *Human friendly mechatronics* (pp. 61–66). Elsevier Science. https://www.sciencedirect.com/sdfe/pdf/download/eid/3-s2.0-B9780444506498500127/first-page-pdf
119. SEBoK. (2024). *Systems engineering heuristics.* https://sebokwiki.org/wiki/Systems_Engineering_Heuristics#:~:text=been%20little%20explored.-,Practical%20Use,process%20has%20to%20be%20accelerated
120. Gligorea, I., Cioca, M., Oancea, R., Gorski, A.-T., Gorski, H., & Tudorache, P. (2024). Adaptive learning using artificial intelligence in e-learning: A literature review. https://www.researchgate.net/publication/376298332_Adaptive_Learning_Using_Artificial_Intelligence_in_e-Learning_A_Literature_Review?__cf_chl_tk=9509_rYODDS0AfWjstQ_cQLJ1i1YTS_oNtslj7iTQbE-1746822659-1.0.1.1-feZSY.4I476Au7ATyrWKecIfJO5lzoZPOn8rnZ2s1VI
121. Khan, K. (2025). Lecture notes on modern software engineering: Principles and practices. https://www.researchgate.net/publication/387795160_Lecture_Notes_on_Modern_Software_Engineering_Principles_and_Practices. https://doi.org/10.13140/RG.2.2.10800.47367
122. Sellevåg, S. R. (2022). Abstraction-decomposition space for critical infrastructure systems: A framework for infrastructure planning and resilience policies. *Security and Defence Quarterly, 39*(3), 6–20. https://securityanddefence.pl/Abstraction-decomposition-space-for-critical-infrastructure-systems-A-framework-for,146789,0,2.html. https://doi.org/10.35467/sdq/146789
123. Pink, D. H. (2009). *The puzzle of motivation.* TED Conferences. https://www.ted.com/talks/dan_pink_the_puzzle_of_motivation
124. Cvet, M. (2023). *Hierarchy and abstraction.* Medium. Retrieved May 9, 2025, from https://mikecvet.medium.com/hierarchy-and-abstraction-3bbde886b41e

125. 8th Light. (2024). *How effective abstractions are opinionated.* https://8thlight.com/insights/how-effective-abstractions-are-opinionated
126. Myol. (2024). *Does the dependency inversion principle apply within the application?* https://stackoverflow.com/questions/78796242/does-the-dependency-inversion-principle-apply-within-application-layers
127. Reemet. (2021). *Correlation versus dependence versus causality—Cross validated.* https://stats.stackexchange.com/questions/509141/correlation-vs-dependence-vs-causality
128. Gurnov, A. (2024). *Why should I use dependency management in project management?* https://www.wrike.com/project-management-guide/faq/why-should-i-use-dependency-management-in-project-management-software/
129. Parr, K. (2021). *How to manage code dependencies by shimming your abstractions.* https://www.freecodecamp.org/news/manage-code-dependencies-by-shimming-your-abstractions/
130. IBM. (2015). *Dependency relationships in UML modeling.* https://www.ibm.com/docs/en/dmrt/9.5?topic=diagrams-dependency-relationships
131. Sangal, N., Jordan, E., Sinha, V., & Jackson, D. (2005). *Using dependency models to manage complex software architecture.* https://groups.csail.mit.edu/sdg/pubs/2005/oopsla05-dsm.pdf
132. Suzín, N. (2024). *Separation of concerns.* Medium. https://medium.com/@natsuzin22/separation-of-concerns-27155f1822fc
133. Parnas, D. L. (1972). On the criteria to be used in decomposing systems into modules. *Communications of the ACM, 15*(12), 1053–1058. https://doi.org/10.1145/361598.361623
134. Beck, K., Beedle, M., Van Bennekum, A., Cockburn, A., Cunningham, W., Fowler, M., Grenning, J., Highsmith, J., Hunt, A., Jeffries, R., Kern, J., Thomas, D., et al. (2001). *Manifesto for Agile software development.* https://agilemanifesto.org
135. Vedal, H., Stray, V., Berntzen, M., & Moe, N. B. (2021). *Managing dependencies in large-scale agile.* Department of Informatics, University of Oslo. https://www.researchgate.net/publication/355489164_Managing_Dependencies_in_Large-Scale_Agile
136. Bailey, B. (2023). An entangled heterarchy. Semiconductor engineering. https://semiengineering.com/an-entangled-heterarchy/
137. Woods, R. (2024). *Software system architecture.* https://www.viewpoints-and-perspectives.info/home/viewpoints/?utm_source=chatgpt.com
138. Views, Models, & Architectures. (2024). *MBSE granularity.* https://caminao.blog/2017/05/27/mbse-granularity/?utm_source=chatgpt.com
139. Osis, J., Nazaruka, E., Donins, U., & García Díaz, V. (2014). Dependencies among architectural views got from software requirements based on a formal model. *Applied Computer Systems, 16*(1). https://www.researchgate.net/publication/276511474_Dependencies_among_Architectural_Views_Got_from_Software_Requirements_Based_on_a_Formal_Model
140. System Architecture. (2024). *Untangle complex system architecture dependencies.* LinkedIn. https://www.linkedin.com/advice/0/how-would-you-untangle-complex-dependencies-ufuhf
141. INCOSE. (2018). *System impact analysis: A systems engineering best practice for change management.* https://www.incose.org/docs/default-source/working-groups/healthcare/public-library/2018-se-in-healthcare-conference/system-impact-analysis-se-best-practice-for-change-management.pdf
142. Ocket, N. (2023). *Clean architecture: The essence of the dependency rule.* Medium. Retrieved May 10, 2025, from https://medium.com/@aboutcoding/clean-architecture-the-essence-of-the-dependency-rule-969f1e8417f6
143. Ingram, D. (2021, August 18). *Best practices for dependency management.* Google Cloud Blog. https://cloud.google.com/blog/topics/developers-practitioners/best-practices-dependency-management

144. Salado, A. (2021). A systems-theoretic articulation of stakeholder needs and system requirements. *Systems Engineering, 24*(2), 83–99.
145. The Metrics Guy. (2025). *Navigating the problem space vs. solution space dilemma.* Nalpeiron. https://nalpeiron.com/blog/problem-space-vs-solution-space
146. Joglekar, N. R., Yassine, A. A., Eppinger, S. D., & Whitney, D. E. (2001). Performance of coupled product development activities with a deadline. *Management Science, 47*(12), 1605–1620. https://www.researchgate.net/publication/264372329_Performance_of_Coupled_Product_Development_Activities_with_a_Deadline
147. He, Z., & Zhu, P. (2008). *A note on multi-response robust parameter optimization based on RSM*. https://doi.org/10.1109/icmit.2008.4654526
148. Fiveable. (2024). *Key concepts of abstraction in programming to know for AP computer science principles.* https://fiveable.me/lists/key-concepts-of-abstraction-in-programming
149. McKenna, R. (2024). *Functional vs. imperative programming.* https://www.ryanhmckenna.com/2014/11/functional-vs-imperative-programming
150. Te'eni, D., & Sani-Kuperberg, Z. (2005). Levels of abstraction in designs of human–computer interaction: The case of e-mail. *Computers in Human Behavior, 21*(5), 817–830. https://doi.org/10.1016/j.chb.2004.02.028
151. Lanford, G. O. (2012). *Engineering system integration: Theory, metrics, and methods* (p. 272). CRC Press.
152. Hehn, J., & Mendez, D. (2021, December 10). *Combining design thinking and software requirements engineering to create human-centered software-intensive systems.* arXiv. https://doi.org/10.48550/arXiv.2112.05549
153. Li, Y., Du, T., Srinivasan, S. G., Wu, K., Zhu, B., Sifakis, E., & Matusik, W. (2022). *Fluidic topology optimization with an anisotropic mixture model.* https://arxiv.org/abs/2209.10736
154. Blanchard, B. S., & Blyler, J. E. (2016). *System engineering management* (pp. 28–32, 242). John Wiley & Sons, Inc.
155. Shishko, R., Aster, R. W., Chamberlain, R. G., McDuffee, P., Pieniazek, L., Rowell, T., Bain, B., Cox, R. I., Mooz, H., & Polaski, L. (1995). *NASA systems engineering handbook* (pp. 70). http://hdl.handle.net/2060/19960002194
156. Endsley, M. R. (2017). From here to autonomy: Lessons learned from human–automation research. *Human Factors, 59*(1), 5–27.
157. Sheridan, T. B. (2002). *Humans and automation: System design and research issues.* Wiley.
158. Booher, H. R. (Ed.). (2003). *Handbook of human systems integration* (p. 70). Wiley.
159. ISO/IEC 25010. (2011). Systems and software engineering—Systems and software Quality Requirements and Evaluation (SQuaRE)—System and software quality models.
160. Norman, D. A. (2013). *The design of everyday things* (Revised and Expanded ed., p. 31). Basic Books.
161. Wickens, C. D., Hollands, J. G., Banbury, S., & Parasuraman, R. (2015). *Engineering psychology and human performance* (4th ed., p. 124). Psychology Press.
162. Lee, J. D., & See, K. A. (2004). Trust in automation: Designing for appropriate reliance. *Human Factors, 46*(1), 50–80.
163. Parasuraman, R., & Riley, V. (1997). Humans and automation: Use, misuse, disuse, abuse. *Human Factors, 39*(2), 230–253.
164. Sarter, N. B., Woods, D. D., & Billings, C. E. (1997). Automation surprises. In G. Salvendy (Ed.), *Handbook of human factors and ergonomics* (2nd ed., pp. 1926–1943). Wiley.
165. Gunning, D., & Aha, D. (2019). DARPA's explainable artificial intelligence program. *AI Magazine, 40*(2), 44–58.
166. Doshi-Velez, F., & Kim, B. (2017). Towards a rigorous science of interpretable machine learning. arXiv preprint arXiv:1702.08608

167. Lyons, J. B. (2013). Being transparent about transparency: A model for human–robot interaction. *Human Factors, 55*(3), 521–534.
168. Kochenderfer, M. J. (2015). *Decision making under uncertainty: Theory and application* (p. 66). MIT Press.
169. Checkland, P. (1999). *Systems thinking, systems practice* (p. 89). Wiley.

The Discovery of Organization

2

Discovery and organization are critical to supporting the systems engineering process during project development. In complex projects, the data involved is often large in volume, abstract, and not fully defined at the outset. Despite this uncertainty, teams must gather, evaluate, and analyze information to keep the project on track. According to the San José State University School of Information [1]:

- Discovery refers to the evaluation and use of information.
- Organization involves describing and classifying data from various sources and disciplines.

The method of obtaining this data may vary. Still, the process of organization and discovery involves identifying, cataloging, and classifying information to make it helpful, understandable, and supportive of the development task. The usefulness lies in understanding the system's context, shareholders' requirements, and potential issues that may arise, such as researching different approaches, technologies, and architectures that can provide relevant information about the system, its environment, and interactions with other systems. This will uncover patterns, trends, and insights that can inform system design and development. Sometimes, leveraging cross-disciplinary insights can be used as an example to model features or functions, delivering new nuances to the development that were not evident before.

Organization in systems engineering focuses on structuring the analysis to define the system's components, interfaces, and relationships, creating a causal flow—basically, the system's logical and hierarchical relationships, along with their associated dependencies and constraints.

F. A. Macias, *Project Development Simplified*, Synthesis Lectures on Engineering, Science, and Technology, https://doi.org/10.1007/978-3-032-01043-8_2

A well-organized system facilitates clear communication and collaboration among different teams and stakeholders. Organization helps to foster a holistic understanding of the system and its interactions with its environment [2].

The discovery process helps characterize the system's behavior. It is about individual system elements, understanding their interdependencies, and synthesizing information to describe the operational context in which the system will function [3]. This involves collecting and evaluating data to uncover patterns and relationships that frame the system's hierarchical relationships, beneficial for establishing system boundaries and supporting the analysis of the following activities involved in the discovery process:

- Recognizing constraints: Involves identifying limitations imposed by technology, stakeholders, or available resources, including budgetary and temporal constraints [4].
- Revealing Dependencies: Planning the interactions and relationships between system components and identifying how stakeholders influence or depend on various system functionalities [4].
- Facilitating innovation: Using information from other disciplines to identify different perspectives of proposed system development by comparing approaches or solutions from other industries or disciplines that can be mapped to describe system functionality. System and product design now incorporates the application of numerous innovations [5].
- Uncovering requirements: Involves identifying both functional and non-functional requirements as a basis for effective project planning. Functional requirements specify what the system must do, including its features and behaviors. Non-functional requirements define the constraints and quality attributes the system must satisfy, such as performance, security, usability, and scalability [5].

Systems engineering processes benefit from integrating domain-specific knowledge with systems thinking, acknowledging that while domain knowledge is essential, systems reasoning is critical for establishing a comprehensive understanding of these processes, reflecting both aspects. It is vital to recognize the importance of emergent effects in determining the global characteristics of systems, emphasizing that traditional views of managing software or computerized portions in isolation are no longer adequate [6].

The interest in discovery and organization lies in the research process, examining how these concepts are managed across disciplines such as medicine, transportation, and manufacturing. This exploration aims to extract insights applicable to abstract ideas beneficial for new systems development. For example, drawing inspiration from biological systems, particularly DNA's architecture, provides a novel perspective for systems engineering and the development of complex systems [7].

Consider an example of discovering and organizing information through a biological lens by analyzing how DNA and its four bases—adenine (A), thymine (T), guanine (G), and cytosine (C)—can be applied metaphorically or structurally in systems engineering.

The information encoded within these four bases (A, T, G, C) provides a compelling analogy to binary code in computing or symbols in language, demonstrating how intricate instructions can be constructed from a limited set of symbols [7].

This analogy extends beyond mere representation, delving into the processes of discovery and organization inherent in biological and engineered systems. Genes, as specific sequences within DNA, dictate unique functions, mirroring the modularity and functional specificity sought in systems engineering [8]. The discovery of meaningful patterns within genomic sequences parallels the data mining and pattern recognition processes important in engineering, suggesting that similar principles of organization and structure underlie both domains [9].

In this application, DNA is not just a biological model but an information architecture at a molecular level in which:

- A, T, G, C are like binary code in computing or symbols in language.

These four bases encode instructions for building proteins, similar to how software code instructs a computer on its operations. This leads to the identification of meaningful sequences derived from a limited set of symbols (e.g., A, T, G, C).

In fact, genes are specific sequences that serve unique functions. Discovery in genomics involves identifying patterns that define sequences of critical importance, similar to recognizing patterns in data mining applications or engineering system behaviors. This concept can be summarized by the following sequence:

$$\textit{Sequence of bases} > \textit{amino acids} > \textit{protein structures} > \textit{functions}$$

The organization in this example is hierarchical and modular. The rationale is to simplify complexity by grouping related elements, clarifying how they interact and connect, and creating a layered description of the system; in this case, a protein, assembled through the pairing of bases, forms the DNA, which is a system. This has redundant coding and self-correct mechanisms such as base-pair checks and repair enzymes, which in engineering can be mapped to a redundant system, as well as error detection and fault tolerance [10]. The importance of using DNA as an example of discovery and organization is that it organizes information such as:

$$\textit{Information} = \textit{Function} + \textit{Structure}$$

Though this is not intended to be an analysis of DNA, it serves as an example to examine the dynamics of systems, specifically how organizing followed by the subsequent discovery of the system framework, functions, and interrelationships can support planning and conceptualization as applied as follows [10] (Table 2.1):

Table 2.1 Applying DNA [10]

DNA concept	Parallel to systems engineering
Four base code (A, T, G, C)	Instruction sets, design languages, minimal rule-based logic
Gene sequences	Modules or components with specific, testable functions
Transcription/translation	Compiling design into executable systems
Protein folding or structure	Systems architecture emerging from base design
Redundancy/error correction	Fault-tolerant systems, backups, recovery protocols

2.1 Concepts to Systems Engineering

Organizing complex systems effectively helps developers reduce ambiguity and manage complexity throughout the design, development, and improvement process. Several organizational strategies support successful system development:

- Hierarchy formation: Structuring subsystems into layers helps manage complexity by separating what the system must do from how it does it.
- Uncovering requirements: Clear identification of functional and non-functional requirements is essential. Functional requirements define what the system should do, like processing input or managing data, while non-functional requirements describe how it should perform, including speed, scalability, and usability. Together, they guide effective planning and implementation.
- Defining roles and boundaries: Assigning specific functions and scopes to components clarifies responsibilities and interactions, reducing overlap and confusion.
- Standardizing representation: Visual tools like UML diagrams, flowcharts, and data flow diagrams provide a clear overview of system interactions. These models support better communication and decision-making across stakeholders by simplifying complexity.
- Facilitating modularity: Designing systems as collections of self-contained modules allows for easier maintenance and scaling, as changes to one module don't disrupt the entire system.

These principles are foundational to systems engineering and software architecture [11]. Hierarchical structuring or layering, if presented as top-down control, tends to weaken systems and stifle growth rather than bottom-up efforts, especially in discovery, which is seen as thriving under pressure and unpredictability. Real breakthroughs come not from rigid planning or formal education but from trial and error, bold risks, and hands-on experimentation [12].

This strongly links to both discovery and organization, but in a very specific way:

- For discovery, This suggests that innovation happens through messy, real-world experimenting and not through centralized planning or formal instruction. It's about doing, not just thinking.
- For the organization, it critiques rigid, top-down structures. These tend to resist change, smother new ideas, and block growth. In contrast, bottom-up systems are considered as more flexible, decentralized, and adaptive, as per Nicholas Taleb are better suited to discovering and implementing new things.

Perhaps the argument is that discovery thrives in environments where the organization allows for uncertainty, iteration, and individual initiative. Organization through the use of systems engineering provides the necessary structure for development, real progress depends on evolving beyond that framework. It is through developing our own heuristics, personal experience-based rules, that the process can be adapted to solve problems, and push innovation forward.

Discovery is the process of uncovering a system's essential functions, behaviors, and responsibilities. It answers the question: What needs to be abstracted? Developers discover core components, relationships, and patterns that shape the system's structure. This step ensures that abstraction is purposeful, grounded in real system needs, and aligned with user and business goals. Without discovery, abstraction risks being arbitrary or misaligned [13]. In dynamic and uncertain environments, predicting Stakeholder needs and preferences can be difficult; therefore, continuous improvement and design thinking approaches for discovery become even more important [14].

For example, in systems integration, understanding the path to integration relies heavily on knowing the system's underlying details. This understanding is crucial to the discovery process. Effective discovery goes beyond surface-level observation; it requires a multifaceted approach that includes user research, system analysis, and domain expertise. By examining the system from various angles, teams can uncover critical functions, interactions, and constraints. This holistic approach ensures that all relevant factors are considered, leading to a well-informed understanding of what should be abstracted. Without a thorough discovery phase, abstraction can become superficial, obscuring necessary complexity or adding unnecessary layers that complicate the system [15].

Abstraction isn't just simplification; it's a restructuring of how a system is represented. It involves redefining existing schemas, models, or frameworks by extracting shared features from specific instances and combining them into a more general representation. This new schema has fewer details but captures the essential ideas that persist across cases. It balances what is removed with what must remain, allowing developers to focus on the system's core logic without losing sight of what matters [15].

Discovery gives abstraction its direction. Organization gives it structure. Together, they enable system developers to manage complexity, create flexible architectures, and navigate between high-level vision and low-level detail with clarity.

Abstraction facilitates the analysis of complex systems by focusing on the most critical elements while setting aside less relevant details. This results in simplified representations that support effective planning, development, and system management. For abstraction to be adequate, it must present a meaningful view of the system, emphasizing key relational patterns. These patterns represent the core functions, structures, processes, or interactions that determine how the system should function, how it should be organized, and how it will operate within its target environment once deployed. This foundation enables a more structured, focused, and coherent design process [16]. To facilitate this objective, a system may be defined further in terms of the following basic characteristics [17]:

- Complex Integration of Resources: A system brings together a wide range of resources, including people, materials, equipment, hardware, software, facilities, data, and funding. Large-scale systems like those in aviation or manufacturing often require substantial amounts of these resources to function. Coordinating them effectively is critical, as relying on chance introduces significant risk.
- Existence Within a Hierarchy: Every system operates within a broader context or hierarchy. For example, a server is part of a local network, which may be integrated into a more extensive enterprise infrastructure, connected to cloud services, and ultimately linked to the broader internet. The performance and constraints of these higher-level systems directly influence the server's role and capabilities and must be considered during its design, configuration, and operation.
- Subdivision Into Subsystems and Components: Systems can be divided into smaller subsystems and components, depending on their complexity and purpose. Breaking a system down in this way simplifies requirement allocation and helps with analysis. However, because components interact with one another, they cannot be designed in isolation. A system must be understood both as a collection of parts and as an integrated whole. Developers must analyze the components and their interrelationships before reassembling them into a cohesive, functioning system.
- Defined Purpose and Functionality: A system must serve a clear purpose, addressing a specific need effectively and efficiently. It should be capable of achieving its goals, even when faced with competing objectives or constraints from the broader system in which it operates. The system must be designed to fulfill its role in the most effective way possible.

For instance, the discovery phase may involve identifying user needs and technical constraints when developing software for a specific application. These insights are then abstracted into system requirements. Organizing and categorizing this information contributes to a structured framework. This process supports abstraction by reducing complexity, grouping related elements, setting priorities, and clarifying the relationships between them.

In more complex engineering systems, organization often involves defining subsystems, mapping interdependencies, and specifying interfaces. Identifying each subsystem's function and how it connects with others ensures that, despite the system's complexity, it operates as a cohesive and coordinated whole. In many cases, a system is just one part of a larger, more complex system, and its design must take into account that broader context. The essence of systems engineering lies in its capacity to address the intricacies inherent in designing a business function, offering a structured approach that has matured in analyzing and working with complexities [18].

Traditional systems engineering activities are tailored to each organization, task, and phase of the life cycle, necessitating an integrated team approach where different participating organizations share responsibility for fulfilling system engineering objectives [17]. Practical system engineering involves using modeling tools to assist in understanding challenges, verifying operability, and simulating towards an improved design [18].

Defining the organization of a system often requires identifying its key subsystems, their specific functions, and how they connect or interface with each other. This structured approach helps ensure that, even with many moving parts, the system operates as a unified and coordinated whole. The systems engineering approach aligns with reductionist theory by breaking down complex systems into simpler, more manageable sub-problems that can be analyzed and solved individually [19].

Systems engineering applies a structured approach to developing complex products and processes. It draws on principles of reductionism, implying that complex problems can be better understood by breaking them into smaller, manageable parts. This hierarchical thinking clarifies the relationships between system elements, their functions, and user behavior.

While discovery, organization, and abstraction often occur intuitively, managing complexity requires making these methods explicit. Discovery identifies the goals. Organization translates them into actionable steps. Abstraction keeps focus on the most critical aspects, filtering out less relevant details. Treating discovery, organization, and abstraction as interdependent and structured processes, instead of relying solely on intuition, is essential for effective planning and analysis in systems engineering and complex systems development. The evolution of systems engineering involves a continuous adaptation of these interdependent processes to meet the changing demands of complex systems development.

Effective systems engineering requires not only technical skills but also an understanding of the broader organizational and operational context in which the system operates [20]. Systems engineering education must engage the students in both the cognitive domain, developing the ability to perform the techniques, and in the affective domain, transforming the student's belief to recognize the positive value of the systems engineering method [21].

Abstraction helps manage complexity by highlighting patterns and key concepts. It lets people and systems ignore irrelevant details and focus on what matters—what worked, what didn't, and why. The ability to focus in this way leads to useful generalizations, making it easier to keep what's relevant and filter out the noise [22, 23].

To apply knowledge later, system developers depend on abstraction to organize what they've learned [24]. This interplay between discovery and organization, facilitated by abstraction, is essential for adaptive learning and problem-solving in diverse contexts [25]. Consider the following examples, which illustrate the interplay between discovery and organization through abstraction and highlight their role in adaptive learning and problem-solving:

- In personal experience: Reflecting on a challenging group project, an individual might move past specific conflicts and instead abstract a broader lesson, such as the value of collaborative planning, including clear roles, timelines, and communication. Discovery occurs through recognizing that effective collaboration was the key takeaway. Organization happens as this insight is distilled into a general principle that can guide future teamwork beyond the specifics of the original project.
- In system development: A machine learning algorithm processes vast amounts of data, identifying patterns, like relationships between variables that drive future predictions. Discovery is reflected in the algorithm's ability to find meaningful patterns. Organization is evident in how it compresses this learning into a model, enabling predictions without retaining all raw data.

These examples demonstrate how abstraction facilitates the connection between discovery and organization, supporting adaptive learning and effective problem-solving across diverse contexts. The connection between abstraction, discovery, and organization becomes especially clear in research and knowledge creation. Qualitative research reflects this in how researchers approach analysis, often starting without a fixed structure or predefined framework. Faced with a large volume of detailed data, the researcher must sift through it without initially knowing what will prove meaningful.

As patterns and themes begin to emerge, the researcher discovers insights and gradually shapes them into a coherent explanation or theory. This process relies on abstraction to distill complex information into essential ideas and on organization to structure those ideas into a clear, intelligible narrative. Together, these elements drive the transformation of raw data into meaningful understanding [26].

This approach to qualitative research enhances systems engineering by providing tools to understand complex, human-centered aspects that are difficult to capture through quantitative data alone. Similar to how qualitative researchers interpret unstructured information to uncover meaningful patterns, systems engineers frequently encounter ambiguity and incomplete inputs during system design and analysis. The ability to extract key insights from rich, context-dependent information parallels the process of modeling and organizing complex systems.

Discovery facilitates the identification of user needs, constraints, or emergent behaviors, while organization aids in translating those findings into system requirements, functional architectures, or design decisions. By employing these qualitative methods, systems engineering can more effectively address socio-technical dynamics, user behavior, and contextual factors influencing system performance and adoption. When faced with complex or contradictory data, the skill that matters most is the ability to extract key insights [27].

This means analyzing details to uncover the patterns and concepts that explain what's happening. These can then be organized into a coherent framework [28]. The abstraction capability is also critical for deep learning [29]. For example, reading is an abstraction of the visual processing of information from a book. Similarly, intentions, emotions, interpretations, and questions are all forms of abstraction. These abstractions shape how knowledge is organized, offering a framework for building structured understanding and explanations [15].

In complex systems, abstraction organizes components into layers representing specific functions or behaviors. These layers simplify understanding, isolate interactions, and support iterative development without losing system coherence. This structure enables developers to focus on and refine each layer independently, while maintaining the overall integrity.

Abstraction subsequently informs the organization process, as these distilled insights provide a framework for structuring knowledge and building integrated system designs. For example, abstraction can emerge in software development through programming examples, modular design, and overall system architecture.

At each level, software developers may rely on abstractions such as variables, functions, classes, or entire service endpoints, including network-accessible interfaces that integrate specific capabilities, to partition system functionality and reduce architectural complexity [30]. For developers, the value of abstraction lies in uncovering relevant information—identifying patterns, defining core components, and mapping relationships within the system. This discovery process is typically enabled through the following methods [15]:

- Data Analysis: Involves collecting and examining raw data from experiments, observations, abstractions or simulations to understand system behavior (function). This process includes mapping relationships to see how components interact and depend on one another, isolating core components that drive system behavior, and filtering out what is unnecessary or the noise to focus on essential information.
- Detecting trends, regularities, and anomalies: This helps simplify and clarify a system's dynamics. Identifying patterns or recurring behaviors allows developers to generalize and better understand the system's functions.
- Domain Expertise: Applying expert knowledge to identify key system elements and understand the trade-offs at play helps in understanding the problem domain, allowing

one to clearly understand what the system is intended to do and the environment it needs to operate in.

For abstraction to be effective, it needs the right balance. Two key aspects to manage are granularity and coverage. Granularity refers to the level of detail: fine granularity focuses on specific components or functions, while coarse granularity examines the overall system architecture and how major modules interact. Coverage, on the other hand, pertains to completeness; high coverage includes all critical parts and relationships, while low coverage may overlook some important elements, creating blind spots. Striking the right balance between the two makes abstraction more practical and reliable in system development .

Abstraction isn't just about making things simpler. This approach helps structure the analysis of complex systems by making sense of different forms that abstraction can take. One standard method involves predefined user types or generalized profiles that reflect behavior patterns, needs, or preferences. These might include categories like "first-time users," "power users," "mobile-only users," or "data-sensitive users." These abstractions are typically based on prior research, market analysis, or usage data.

Predefined user types help engineers start discovery by framing early assumptions, focusing observations, and prioritizing key data. These general abstractions guide the transition toward more specific patterns and behaviors as real-world user data accumulates. Abstraction, in this sense, supports an iterative process, starting broad and evolving into fine-grained insights that reflect actual user needs and interactions. These user types act as design guardrails, enabling systems to refine models over time and adapt from static templates to dynamic, data-driven representations tailored to individual users [30].

Organizing related components, defining modular boundaries, and structuring functionality help systems engineers build scalable, adaptable, and relevant systems. Clear assessments of tasks reduce cognitive load and improve planning. Planning is a process of discovering the project itself. Monitoring and control evolve, offering insights into how well the solution takes shape. Tracking functional and performance metrics, such as the frequency of changes, gives developers a sense of effectiveness and system stability. These insights, driven by continuous learning and discovery, inform necessary adjustments to the overall scope [31].

By abstracting a system's technical elements and integrating insights from multiple disciplines, developers can create technically robust models that address real-world complexity. Discovery and organization play a central role in supporting this process. Discovery helps uncover underlying patterns, constraints, and opportunities within the system and its environment. Organization gives structure to these findings, making it easier to define boundaries, isolate modules, and manage complexity. Together, they ensure that abstractions are grounded in real data and structured knowledge rather than assumptions.

In model-based development, especially when design strategies are clear, this pairing allows developers to build models that are not only technically sound but also adaptable and aligned with evolving requirements [17].

Large projects often draw on expertise from multiple disciplines. In these cases, abstraction isn't just a way to simplify but becomes a method for making sense of complexity. Organization and discovery support each other by structuring system components and recognizing patterns that developers can use to create clarity.

At the same time, ongoing exploration helps identify what's most important. Together, they ground abstraction in insight and structure, making it more effective at guiding development. Consider the following as examples of different disciplines that can be used to address abstraction [17]:

- Layered abstraction in computer science helps manage complexity by breaking systems into distinct levels, such as hardware, operating systems, and applications. The same principle applies in systems engineering, where high-level goals are separated from low-level implementation details. For instance, when designing a modern transportation system, engineers might define user needs, such as safety and efficiency, at the top level, while handling infrastructure and vehicle mechanics at lower levels. This kind of separation makes problem-solving more manageable and supports better collaboration across different fields [30, 32].
- Mathematics offers powerful tools for abstraction, using formal representations such as differential equations, state-space models, and set theory to describe real-world systems. In engineering, these techniques enable the creation of models that capture essential system behaviors [33]. Through mathematical formalisms, engineers can predict performance under various conditions, analyze complex interactions, and design systems that are scalable and manageable [34]. Modeling is central to modern systems design. A model serves as an abstraction of a system from a specific viewpoint, describing it at a chosen level of precision [35].
- Biological systems provide a model of modularity, where complex organisms are composed of semi-independent subsystems that function together. This concept has influenced engineering, encouraging the use of modular design to simplify development, testing, and maintenance [36]. In systems like robotics, modular components such as detachable arms or sensors can operate independently while still integrating smoothly into the larger system. This mirrors the adaptability and resilience found in nature, allowing engineered systems to be more flexible, scalable, and easier to evolve over time [37].
- Analogical reasoning plays a key role in human thought, learning, and problem-solving [38]. Drawing from this, cognitive science provides methods for abstracting complex functions through analogy and conceptual models. Engineers often apply this approach when designing systems that replicate human abilities such as vision, reasoning, or language processing [39].

- Neural networks in artificial intelligence exemplify this abstraction, drawing inspiration from biological neural systems. Engineers develop algorithms that can learn and adapt by mimicking the human brain's processing of information, enabling advancements such as wireless systems, autonomous vehicles, and medical diagnostics [40].

In the paper Toward Artificial General Intelligence (AGI) Native Wireless Systems for the Metaverse, the authors present a vision for building wireless systems capable of supporting metaverse applications, which are defined as collective virtual spaces where digital and physical realities converge. These spaces combine technologies such as augmented reality (AR), virtual reality (VR), and 3D environments [41]. The paper proposes a shift from traditional AI to AGI native wireless systems, which are equipped with common sense and advanced cognitive abilities such as perception, reasoning, and analogy, considered essential for adapting to unpredictable and complex scenarios. A proposed network development is defined to be built on three core components [42]:

1. Perception Module—Translates real-world data into abstract, generalizable forms.
2. World Model—A high-dimensional, causal representation grounded in intuitive physics.
3. Action Planner—Uses the world model to make decisions and plan actions, inspired by the human brain.

Abstraction enables AGI-native wireless systems, particularly perception, reasoning, and planning. It also serves as a bridge between discovery and organization, especially in the context of analogical reasoning. In an AGI native system, the perception module does more than gather raw data. It transforms that data into abstract, generalizable representations. Rather than processing every detail, it identifies key patterns and relationships across different situations.

Such abstraction allows the system to recognize underlying similarities between scenarios even when surface details differ. This capability is essential for achieving scalability and generalization in dynamic and unpredictable environments. Abstraction enables the discovery of patterns across diverse data and organizes knowledge into structured forms, supporting flexible reasoning and adaptive decision-making [42].

Analogical reasoning relies on abstraction. To understand a comparison like "an atom is like a solar system," one must move beyond specific details such as protons, electrons, planets, and stars, and instead focus on the shared structure: a central object with orbiting elements. In complex engineered systems, abstraction facilitates the development of internal models that represent entities and their relationships in a manner that enables this kind of reasoning. A system model can be viewed as a high-dimensional, abstract representation of reality. Rather than storing isolated facts, it captures cause-and-effect relationships, system dynamics, and core physical principles. This allows the system to simulate outcomes, predict behavior, and adapt to new situations. In system development,

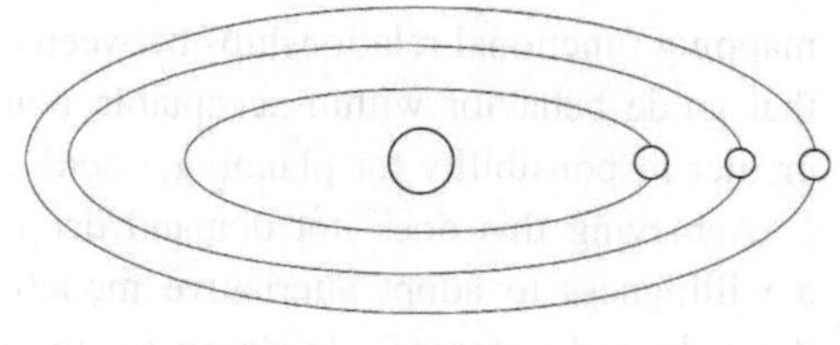

Fig. 2.1 A central object with orbiting elements [43]

it supports flexible, context-aware operation without the need to explicitly program every possible scenario [42] (Fig. 2.1).

The action-planning looks at abstract goals rather than specific instructions. For instance, instead of being told exactly how to respond to an event, the AGI-native system is driven by intent, or rather, to "optimize user experience." This depends on abstract models to strategize for achieving those outcomes in different contexts [42].

Reducing ambiguity benefits from borrowing concepts from other disciplines, such as the wireless-native network example of models, and applying the perception and world model, action planner, and analogical reasoning to address aspects of development. These models provide a structure for analysis and help identify ways to manage both uncertainty and the complexity that comes with it. Focusing on developer experience rather than user experience is more appropriate in this context. The goal is to reduce the cognitive burden on developers working with advanced, open-ended systems by developing frameworks that make the development more straightforward to understand, design, adapt, and maintain.

Borrowing from other disciplines to develop abstractions is not meant to define strict rules but to help developers find ways to model system behavior more effectively. For example, design thinking encourages clarity, modularity, and intent, making breaking complex systems into understandable and testable parts easier. Cognitive science offers models for reasoning and perception that can guide system architecture. Drawing inspiration from design thinking, for example, one can emphasize the importance of clarity, modularity, and a well-defined intent within system architecture, facilitating the decomposition of intricate systems into manageable, comprehensible, and testable components [44].

This method not only simplifies the development process but also encourages a deeper understanding of the system's underlying mechanics, which supports the rationale for finding other ways to enhance the analysis [45]. Using these methods as tools enhances engineers' overall analysis process. The focus shifts from constructing narrowly defined procedural logic to operating at the system level [15]. Procedural logic is a step-by-step instruction sequence that dictates how the system should respond in specific situations [46].

While effective in controlled or straightforward environments, scaling in more complex or unpredictable settings becomes rigid and demanding. Rather than detailing every low-level behavior, the development process prioritizes defining system-level requirements, mapping functional relationships between components, and setting operational constraints that guide behavior within acceptable bounds. This shift enables developers to take on greater responsibility for planning, coordination, and adaptation.

Achieving this does not demand deep expertise in external disciplines, but instead, a willingness to adopt alternative modeling perspectives that simplify complexity and drive desired outcomes. In doing so, the engineering process mirrors the goals of user experience by emphasizing how intuitively and effectively engineers can engage with the system's conceptual structure. Such an approach is particularly valuable in environments marked by uncertainty or evolving requirements. For instance, designing a system that integrates dynamic, real-time perception and planning capabilities benefits from these modeling strategies, streamlining system behavior and enhancing integration with other components or functions [15].

2.2 Abstraction Example Using Medical Diagnostics to Establish a Testing Framework

Consider the example of an autonomous car proposed as a new project. Critical to success is ensuring that the system will operate in its intended operational environment as envisioned. A key consideration is the vehicle's ability to function with minimal or no human supervision, which involves assessing its cognitive capabilities, specifically how well it can adapt to and perform across a wide range of circumstances. In addition, the system must be resilient against external threats such as hacking, signal interference, or other forms of intrusion that could compromise its functionality. The broader challenge lies in ensuring reliable performance across all conditions that may impact safety, stability, or system performance [47].

Testing to ensure that the system performs as needed will be required to determine trust, as the system operates autonomously and makes decisions with little or no human intervention. Determining the necessary metrics to assess biases and the explainability of decisions that support user trust in the vehicle will require its analysis [48]. Deriving needed information to structure the development process by identifying key data inputs will require the discovery and organization of information, which can be structured as follows:

2.2.1 Discovery

The discovery phase focuses on understanding what the system needs to do, uncovering unknowns, and identifying potential risks or gaps in knowledge related to autonomous system performance [48, 49]. Consider the following:

- Identifying system goals: Clarify success by specifying outcomes such as safe navigation, minimal human intervention, real-time decision-making and adaptation, and resilience to unusual or unexpected situations that may fall outside desired performance and operating conditions [50].
- Recognize operational challenges: Understand the range of operating conditions that may impact performance, such as weather, road types, traffic patterns, unexpected obstacles, and interactions with human drivers or pedestrians (always a problem) [51].
- Mapping Out Cognitive Requirements: Identify the perception, reasoning, and planning capabilities necessary for autonomous operation. This includes how the system detects and interprets its environment, prioritizes tasks, makes real-time decisions, and adapts to ambiguity or unfamiliar scenarios [52].
- Recognize Threats: Identify potential vulnerabilities that could compromise performance or safety, including cybersecurity attacks, GPS spoofing, sensor tampering, and other forms of system interference [53].
- Learning from Failure: Collect and analyze data from simulations, test drives, and real-world incidents to identify patterns in system failures, understanding when, where, and why autonomous functions break down [48].

2.2.2 Organization

Once important performance elements are characterized, the next step is organizing that information into a structured framework to address testing and development requirements. Consider the following:

- Structuring System Architecture: Define how core modules—perception, world modeling, planning, and control—interact within the system. Maintaining a clear separation of responsibilities and model interfaces ensures reliable communication between components. Account for dependencies with external systems such as traffic infrastructure, network services, and vehicle-to-vehicle (V2V) communication [54].
- Prioritizing Test Scenarios: Create test cases spanning routine driving situations to rare, high-risk edge cases. Develop scenario libraries to systematically organize these cases, monitor test coverage, and identify gaps. This approach helps ensure that both common functionality and critical safety limits are thoroughly evaluated [55].
- Defining Evaluation Metrics: Establish clear metrics to assess safety, performance, reliability, adaptability, and security. Each system module should be measured against criteria that reflect the operational goals of the autonomous vehicle. Beyond traditional Measures of Performance (MOP) and Measures of Effectiveness (MOE), the metrics should support assessments of system trustworthiness—a more nuanced and multifaceted challenge [56].

- Modeling Interfaces and Dependencies: Define how the autonomous vehicle interacts with external systems such as traffic control infrastructure, cloud-based services, and vehicle-to-vehicle (V2V) communications. This includes specifying data exchange protocols, latency constraints, and failure modes to ensure seamless and safe integration [57].
- Creating a threat response framework: Organize threat models, intrusion detection strategies, and fallback plans into a coherent structure to test security alongside core functionality [58].

Models are typical in the design space; it makes sense to extend the use of models to the requirements activity. Traditionally, prototype models are developed to further analyze, understand, and discover requirements [17].

A significant system engineering challenge is to evolve through the development process using an organized, logical, and methodical approach, and utilizing the available techniques and tools. The goal is to gain early visibility, reduce the time it takes for system acquisition, make sound design and management decisions as system development evolves, and minimize the risks associated with the decision-making process [17].

Discovery reveals system capabilities and potential failure points, while organization turns those insights into a structured foundation for design, testing, and refinement. Both are essential for navigating the ambiguity and complexity of developing autonomous systems. As the project progresses, models related to interfaces, dependencies, and analytical methods evolve to approximate or represent key characteristics such as structure, behavior, operation, and support. This ongoing process of simplifying and concentrating on required system attributes forms the basis of abstraction [17].

Figure 2.3, aims to illustrate discovery as a structured approach to problem-solving that relies on abstraction, synthesis, and implementation processes, which are considered core methodologies in systems engineering and software architecture.

The process begins with identifying the problem, which is the specific challenge or need to be addressed through the realization of a solution. Through abstraction, this problem is generalized to reveal broader patterns or principles, resulting in a generalized problem. This abstraction step must strive to remove unnecessary details, making it easier to identify realistic approaches and core issues.

Synthesis is applied to formulate a generalized solution rather a conceptual framework or methodology that addresses the generalized problem. This stage often involves modeling, analysis, and theoretical analysis.

Implementation bridges the gap between theory and practice by translating the generalized solution back into the specific context. The result is a practical and effective response to the original problem which continues to evolve as implementation and testing uncover new requirements or challenges that must be addressed.

The final implementation phase delivers a functional, context-specific solution, completing the systems engineering cycle. This model underscores the importance of abstraction and synthesis in managing complexity by transforming specific problems into generalized frameworks, then engineering targeted solutions. The approach reflects a disciplined process of defining system boundaries, identifying key variables, and aligning technical design with operational needs to ensure robust, scalable outcomes.

This structured progression, from abstraction to implementation, relies on early discovery to uncover the critical functions, constraints, and failure points that shape effective system solutions (Fig. 2.2).

One effective way to learn something new is through specific examples rather than abstractions and generalities, which help understand the overall complexities of complex systems [59]. Examples can support the development of abstraction. For example, consider the four-box framework, a problem-solving method often used in science and technology. By breaking a problem into clear, explicit parts, the framework helps strip away unnecessary details to reveal the core issue, abstracting the problem into a more general form.

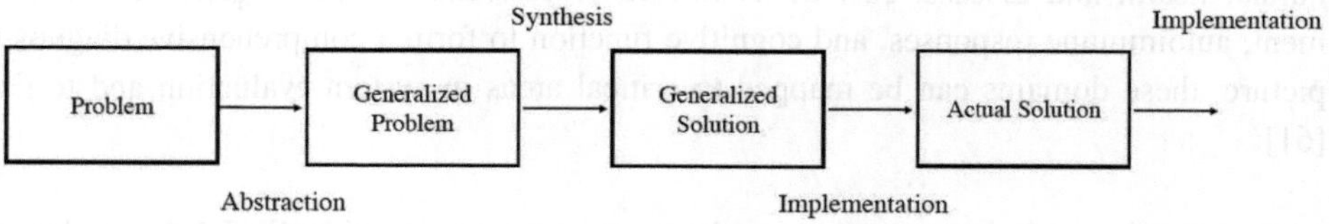

Fig. 2.2 The link of discovery to abstraction to solution [60]

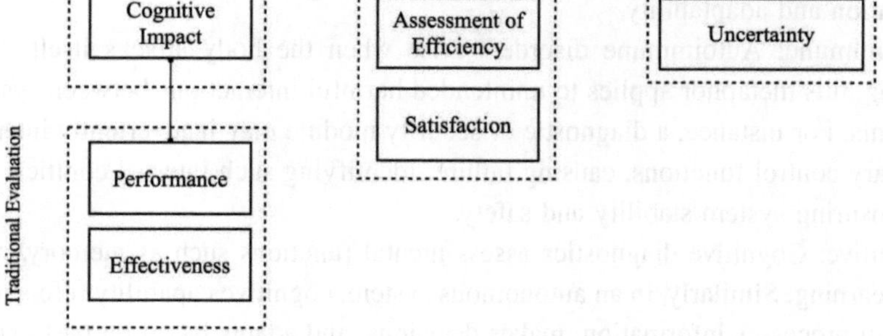

Fig. 2.3 Hierarchical abstraction in systems evaluation [30]

Existing solutions or models then guide the abstraction, helping verify and clarify key attributes. These insights feed back into the analysis, sharpening it with relevant detail. The framework encourages a back-and-forth between abstract and critical thinking, making problem-solving more effective and focused [60].

Abstraction is fundamental in systems engineering for managing complexity and focusing on the essential behaviors of a system. One practical approach is using models and analogies from other disciplines to frame and analyze system development. In this context, reviewing medical diagnostics provides a compelling case study.

Medical diagnostics relies on structured testing, layered decision-making, and iterative refinement to isolate symptoms, confirm hypotheses, and guide treatment, paralleling how complex systems are tested and validated. By mapping diagnostic strategies onto system development processes, engineers can better design test strategies, identify failure modes, and refine system architecture through structured abstraction and targeted analysis.

To effectively test and validate the behavior of an autonomous vehicle, principles can be abstracted from the multifaceted approach used in medical diagnostics to understand human health and disease. Just as healthcare professionals evaluate genetics, environment, autoimmune responses, and cognitive function to form a comprehensive diagnostic picture, these domains can be mapped to critical areas in system evaluation and testing [61]:

- Genetic: In medicine, genetics involves analyzing an individual's DNA to identify predispositions to disease or treatment responses. In system terms, this maps to the foundational architecture and design choices that define the vehicle's baseline capabilities and inherent risks. Understanding a system's "genetic code", its software, hardware, and core configurations can help identify structural vulnerabilities or performance limits.
- Environmental: Just as human health is affected by external factors like pollutants, diet, or social context, an autonomous vehicle's performance is influenced by the operational environment. Road conditions, weather, traffic density, and unexpected hazards are environmental variables that must be tested under diverse scenarios to ensure robust operation and adaptability.
- Autoimmune: Autoimmune disorders arise when the body attacks itself. In system testing, this metaphor applies to unintended harmful interactions between system components. For instance, a diagnostic or security module may inadvertently interfere with primary control functions, causing failure. Identifying such internal conflicts is crucial for ensuring system stability and safety.
- Cognitive: Cognitive diagnostics assess mental functions such as memory, reasoning, and learning. Similarly, in an autonomous system, cognitive capability refers to how the system processes information, makes decisions, and adapts based on past experiences. Testing should evaluate whether the system can perform tasks effectively and improve its performance over time through learning algorithms.

Adapting a medical diagnostic model as an abstraction for testing autonomous systems presents a powerful opportunity to evaluate system behavior, resilience, and adaptability through a more integrated lens. A key advantage of this approach is its focus on domain interdependence. In medicine, genetic predispositions can shape responses to environmental conditions, trigger autoimmune reactions, or influence cognitive development, factors that are not isolated but dynamically linked. This analogy maps well to complex systems, where the foundational architecture (genetic), operating context (environmental), internal feedback and faults (autoimmune), and decision-making logic (cognitive) must be understood as a whole rather than in isolation.

By abstracting system testing through the lens of medical diagnostics, engineers can better identify hidden dependencies, anticipate potential failure modes, and assess adaptive performance in realistic operational conditions. This framework supports a richer, more layered understanding of how autonomous systems behave, particularly when traditional, domain-specific testing may overlook critical cross-domain interactions.

This perspective enables a more integrated analysis of system performance. Evaluating an autonomous system's cognitive capabilities, such as its ability to learn, adapt, and respond within a defined operational environment, can expose hidden dependencies and critical limitations. As an abstraction method, the framework reveals interactions that are often overlooked when system domains are tested in isolation.

By borrowing from these diagnostic analogies, the testing framework becomes a powerful tool for identifying latent vulnerabilities, exposing potential failure modes, and improving the overall robustness and performance of complex, adaptive systems.

2.3 Borrowing from Genetics to Abstract and Test System Functionality

Genetics studies the human system as an intricate composition of interdependent parts: genes, proteins, cells, and organs. This approach's advantage is that it considers how each element supports the whole system, so a disruption in one part can affect the rest. At a high level, genetic analysis identifies inherited traits, mutations, and gene-environment interactions that influence health, forming the basis for targeted diagnosis and intervention.

The systems view in genetics, which focuses on understanding composition, relationships, and potential failure points, provides a strong analogy for systems engineering. Just as genes function within complex biological pathways, engineers can think of systems as networks of interdependent components. In this context, the term inherit describes how a subsystem takes on essential functions, behaviors, or constraints from its parent system. These are not optional but include technical, architectural, or regulatory requirements that new components must incorporate to operate effectively within the larger system.

For example, research in genetics shows that inherited traits, combined with environmental factors, can increase the risk of autoimmune conditions. In much the same way, inherited system constraints can shape how a technical system responds under pressure, highlighting the importance of understanding these built-in dependencies during design and development.

This highlights the interdependencies across domains, just as in engineered systems where performance depends not only on internal design but also on the external environment, which may be shaped by other systems or constraints [62]. By framing a system this way, engineers can better identify what must be tested:

- Inherited constraints must be validated. For instance, does a new subsystem respect the parent system's power or format requirements?
- Interdependent functionality must be tested. If one component's behavior changes, does it disrupt the behavior of others, just as a genetic mutation might alter a biological pathway? [63]
- Environmental factors must be modeled and tested. Systems must be validated against the real-world operational environment, just as genetics considers external factors like toxins or infections that affect health outcomes [64]. The genetic model encourages engineers to consider:
- Map the "genetic" blueprint of the system: what functions and constraints are passed down [65].
- Understand dependency networks: what components must work together, and how changes propagate [66].
- Plan tests systematically: targeting inherited traits, critical interactions, and environmental robustness [66].

2.4 Framework for Testing Using Genetic Abstraction

A testing framework can be modeled after principles in genetic analysis to better understand and validate the behavior of complex systems. Just as genetic diagnostics trace inherited traits, environmental influences, and functional disruptions, this approach abstracts similar concepts such as inheritance, dependencies, and environmental conditions to help guide structured testing in systems engineering. Inheritance refers to prior dependencies within a system, such as legacy interfaces, previously defined performance requirements, architectural constraints, or design decisions from earlier iterations. These inherited elements influence how new components are integrated and must be validated early in the testing process [67].

Consider the following:

- Identify Inherited Elements [65]
  - Define system constraints, dependencies, and properties passed from previous systems or designs (e.g., interfaces, performance metrics, environmental tolerances).
- Map Component Relationships [66]
  - Describe how subsystems interact by identifying their functional relationships, inputs, outputs, and shared resources or data.
  - Focus on critical connections where the behavior of one component can directly influence another, especially in ways that may affect overall system performance or stability.
  - Highlight critical paths where failure or changes could occur.
- Define Functional Inheritance or Dependencies [65]
  - For each subsystem, document what functions are inherited versus newly introduced.
  - Prioritize testing of inherited functionality first for baseline validation.
- Analyze Environmental Dependencies [65]
  - Environmental conditions define what may be experienced in the operational context. For example, electromagnetic interference, temperature fluctuations, and radiation exposure can significantly affect system performance.
  - These factors should be identified and characterized to ensure the system is tested under conditions that reflect real-world operational challenges, ensuring performance or identifying weaknesses.
  - Model environmental variations to stress the system during testing.
- Develop Targeted Test Cases [67].
  - Constraint Testing: Validate that inherited requirements are appropriately managed.
  - Interdependency Testing: Simulate failures or changes in one component to verify system stability.
  - Environmental Stress Testing: Operate the system across realistic and extreme conditions.
- Design refinement based on test outcomes [67]
- Test results can be used to refine system architecture, update component interactions, and adjust inherited requirements.
- The derived feedback process ensures the system evolves in response to real-world performance insights, such as how genetic data informs treatment decisions in medical diagnostics.

2.4.1 Environmental Diagnostics as a Model for System Testing [67]

Medical diagnostics widely recognize environmental factors as key contributors to disease development. Environmental pollutants, air quality, water contamination, noise, and radiation are monitored to understand how external exposures may affect health. Medical diagnosis often focuses on how these conditions trigger or worsen the disease, informing more targeted testing and treatment approaches [68].

Applying this approach to systems engineering, environmental abstraction becomes a method for aligning system performance with real-world operational conditions. Just as medical diagnostics account for the external conditions affecting human health, systems engineering must translate environmental variables into design and test parameters [68].

For example, integrating a wireless communication module into a vehicle requires detailed modeling of the spectral environment. The system must remain reliable despite fluctuations in the noise floor, caused by background radio frequency (RF) energy, nearby transmitters, or other forms of interference. These variations must be anticipated and factored into environmental stress tests [69].

A concept in biology suggests that a properly functioning immune system monitors its environment, responds appropriately to external threats, and maintains internal stability. When this environmental sensing breaks down, the body becomes vulnerable to infection, inflammation, or broader systemic stress. Similarly, engineered systems must be capable of detecting and distinguishing between normal operating conditions and those altered by disruptive environmental factors [62].

While a close focus on system requirements is fundamental to development, the added value of abstracting functional behavior and environmental context can reveal system vulnerabilities that might otherwise go unnoticed. Using a diagnostic model to structure this analysis provides a tool to focus the analysis on identifying areas where performance may degrade. Overlooking these distinctions can reduce system effectiveness or compromise its intended function, highlighting how borrowing from other disciplines broadens and sharpens our understanding during system development. For example, considering how the immune system responds to environmental factors can offer insights into system behavior. In an engineering context, this could be seen in how electromagnetic interference, representing the environment, impacts a communication subsystem's ability to function reliably, much like environmental stressors affect immune response in biology.

Reliable system performance depends on understanding the environmental conditions that may influence functionality. By modeling normal and extreme operating environments, engineers can identify how those conditions might challenge the system. This enables the discovery of potential points of failure or instability and supports the development of targeted tests to evaluate the system's capacity to withstand and adapt to environmental stress [70].

Borrowing from the medical diagnostic model, the connections between domains such as how environmental conditions influence genetic predispositions these can significantly impact outcomes. For example, a dormant condition might become active only when triggered by an external stimulus. Similarly, a system's internal behavior can change depending on specific conditions in its operational environment.

This diagnostic perspective supports discovery by helping engineers identify underlying system influences and supports organization by structuring those influences across relevant domains. Whole system analysis, incorporating environmental, genetic, autoimmune, or cognitive factors, contributes to a more comprehensive understanding of system function. Applying this approach in systems engineering enables abstraction in development and helps anticipate how contextual factors may affect performance. Consider the following [62]:

Framework for Environment-Based System Testing

- Identify Environmental Variables
  - List relevant environmental factors for the system (e.g., noise, vibration, radiation, temperature, humidity, RF interference).
- Characterize Baseline and Extreme Conditions
  - Define typical operational environments and potential worst-case scenarios.
- Model Environmental Inputs
  - Create simulations or physical test setups that expose the system to expected and extreme environmental conditions.
- Analyze System Vulnerabilities
  - Predict how environmental factors might impact system performance (e.g., signal degradation under RF noise).
- Design Environmental Stress Tests
  - Environmental Condition Testing: Validate system function under typical environmental loads.
  - Stress Testing: Push system performance under worst-case environmental scenarios to uncover hidden weaknesses.
- Develop Discrimination Criteria
  - Ensure the system can recognize and appropriately respond to environmental changes (e.g., adaptive filtering in RF systems).
  - Analogous to how the immune system distinguishes between self and non-self.
- Feedback and Adaptation Loop
  - Testing data to refine system design makes it more robust to environmental challenges.

This approach encourages engineers to think diagnostically by asking what environmental factors could cause the system to fail, how they can be modeled, and what tests are needed to demonstrate the system's ability to withstand or adapt to them.

2.5 Abstracting Immunity to Identify and Counter Challenges to System Operation

The immune system exemplifies a model that helps us understand its biological attributes, which aids in engineering resilient systems. Its layered defense mechanisms and innate and adaptive responses illustrate how systems can detect, respond to, and adapt to a variety of challenges while maintaining stability. Its modular structure, feedback control, and capacity to learn from past exposures offer valuable design principles for developing resilient, adaptive technologies [71].

This combination of resilience and adaptability in the immune system can be abstracted into a systems development approach to guide the design of systems that maintain functionality under stress, attack, or failure. Immune-inspired design principles can enhance system function by enabling: [72]

- Recognize and respond to abnormal internal behavior, such as sensor faults, degraded performance, or corrupted data;
- Identify, characterize, and mitigate external threats, including cyberattacks, environmental hazards, or adversarial interference;
- Recover from disruptions and adapt operational strategies in real time without external intervention or a complete system reset.

For example, consider a communication network facing a potential cyberattack. An "immune" response in this context would extend beyond external firewalls to include internal anomaly detection, self-healing protocols, and dynamic reconfiguration to maintain continuity of service. This reflects how biological systems detect, respond to, and recover from infections to restore internal balance [73].

Biological systems are inherently complex, adaptive, and stochastic. Replicating these traits into a systems development process can pose significant challenges, especially modeling their dynamic behavior and maintaining consistent, reliable performance, which is the purpose of the modeling. However, applying biological abstraction to interpret and implement functionally requires translating biological insights into precise engineering specifications. The opportunity lies in creating modular components that mimic biological functions to develop models to help analyze and predict system behavior. Achieving this demands close collaboration across disciplines, standardized design frameworks, and rigorous validation protocols to ensure functionality and reliability. The complexity of the analysis depends on how much detail the abstraction needs to capture, which varies based on the type of insight or prediction the analysis aims to produce [74].

Continuing the development of the framework, the next step draws inspiration from biological immune systems to define, simulate, and test threats that could compromise system performance. It addresses both internal and external disruptions by abstracting threat types, identifying patterns of abnormal behavior, and automating appropriate responses.

The framework also incorporates recovery modeling to evaluate system resilience and integrates with systems engineering tools to support continuous validation and design adaptation. By mirroring how biological systems detect and respond to threats, this approach provides a structured method for enhancing robustness and adaptability in complex engineered systems [17, 75, 76]. Consider the following:

- Threat Abstraction: External threats: cyber intrusions, electromagnetic interference, unauthorized access, environmental disruptions.
  - Internal threats: corrupted states, failed components, misconfigurations, software bugs.
- Pattern Recognition
  - Develop detection models to identify threat signatures.
  - Use AI/ML to recognize anomalies and learn evolving threats, akin to immune memory.
- Response Mechanism Design
  - Implement multi-layered defenses: firewalls, sandboxing, dynamic resource isolation.
  - Develop automatic response protocols: component isolation, rollback, reallocation of functionality.
- Recovery Modeling
  - Design self-healing systems: auto-correction, redundancy, adaptive routing.
  - Integrate fail-safe mechanisms that protect critical functions under attack.
- Digital Twin Integration
  - Use Digital Twins to simulate immune-like responses under attack scenarios.
  - Run stress simulations to validate system robustness under diverse and compounded threats.
- Model-Based Systems Engineering (MBSE)
  - Structure all threat-response-recovery mechanisms within an MBSE framework.
  - Ensure traceability, testability, and cross-domain coherence of "immunity" features.
- Key Technologies Supporting Immunity Abstraction
  - AI/ML: Enables adaptive pattern recognition and real-time threat classification.
  - Digital Twins: Support continuous testing and modeling of system behavior.
  - Model-Based Systems Engineering (MBSE): Unifies requirements, behavior modeling, and threat response validation.

By modeling system threats after the core functions of the immune system, engineers can better identify vulnerabilities, predict failure modes, and design for resilience. Studying how biological immunity adapts and recovers in real time helps define fault-tolerant architectures that maintain performance under stress or disruption.

Cognitive Properties as a Model for Intelligent System Design and Test

In humans, cognition is essential for understanding the world, encompassing perception, processing, learning, and action. These cognitive processes form the basis of human intelligence, enabling the interpretation of surroundings, pattern recognition, problem-solving, and behavioral adaptation based on learning.

Learning extends beyond simple data storage, enriched by experiences, values, and context, which supports rationalization, adaptation, evolution, and intellectual and emotional functioning [77]. In both clinical and technical diagnostics, this cognitive framework is crucial, where symptoms or anomalies are perceived, potential causes are reasoned, and decisions are adjusted based on feedback. As AI systems advance, their architectures are increasingly reflective of this model of cognition [78].

In AI applications, the cognitive domain is used to evaluate a system's ability to perceive, process, and learn. Which are defined as follows [79–81]:

- Perception: Perception encompasses the acquisition of raw data, whether it be from sensor readings, textual input, or visual imagery
- Processing: Transforming data into actionable information via rules, patterns, or inference.
- Learning: The system can adapt based on outcomes, feedback, or changing inputs.

Unlike humans, AI lacks true understanding, self-awareness, and ethical reasoning. Its algorithmic cognition is limited to patterns on which it has been trained. Without human input, it can process data but not meaning, intent, or consequences. However, this limitation can also be its strength, which is that AI excels at analyzing vast data and detecting complex patterns, but it needs external guidance to reason about context and implications [82].

Still, the gap is narrowing. With advances in large language models, AI is increasingly able to simulate context-sensitive reasoning and generate responses that appear meaningful. This can mimic ethical reasoning by drawing from human-trained data and aligning outputs with predefined goals or values. While this isn't true understanding, it shows how structured guidance and training can push AI closer to human-like judgment in specific tasks [83].

In clinical and technical diagnostics, the cognitive model is useful since it drives decision-making, for example, how to perceive anomalies, reason about possible causes, and refine our actions based on feedback. This loop of observation, inference, and adaptation is central to human problem-solving and increasingly embedded in advanced AI systems. As these systems evolve, they are designed to mirror this model by detecting irregularities, hypothesizing faults, and learning from outcomes. This shows the benefit of adapting the functionality of a cognitive model to abstract how an AI-enabled or autonomous system will function, or more importantly, when it will not [84].

At this point, revisiting the Hierarchical Abstraction in Systems Evaluation model provides a useful way to reframe practical evaluation through a systems engineering lens. Although the model was first applied in healthcare, its core ideas, such as measuring real-world effectiveness, usability, and how well a system meets user goals, are relevant to complex systems in many fields. These concepts support a structured and purpose-driven approach to evaluating performance in real operational settings (Fig. 2.3).

This perspective applies not only to human systems, such as clinicians making decisions, but also to technical systems like AI-driven tools or autonomous platforms. Framing evaluation through hierarchical abstraction enables the model to adapt to systems engineering and other technical domains where human–machine interaction and informed decision-making are crucial to system performance [84, 85]. This method supports the discovery and organization of information, clarifies system functionality, and reveals the layered relationships that influence behavior, making it particularly useful for analyzing complex and adaptive systems [84, 85].

Adapting a medical diagnostic evaluation framework provides a structured method for assessing complex systems in a way that mirrors clinical reasoning. This approach begins with examining system outputs, including explanatory components, and extends to evaluating the system's perceived utility in an operational context. Similar to how clinical models are assessed for the accuracy of individual predictions and their overall value in supporting decision-making across diverse scenarios, this framework captures both technical performance and practical relevance.

Borrowing from the medical diagnostic evaluation framework, the benefit of this process is that it provides a structured way to assess the cognitive properties of complex systems in a manner that mirrors clinical reasoning. This approach begins by examining system outputs, including explanatory components, and extends to evaluating the system's perceived utility in an operational context. Just as clinical models are reviewed for the accuracy of individual predictions and their overall value in supporting decision-making across diverse scenarios, this framework considers both technical performance and practical relevance, which are critical in system development [84].

Pragmatic evaluation goes beyond performance metrics by focusing on usability, cognitive impact, and user trust. Cognitive impact refers to how system design affects users' understanding, decision-making, and mental workload. Traditional evaluation methods often overlook these factors, yet they are important for assessing complex, adaptive systems. In these environments, how users interact, the context of use, and decision-making processes directly influence effectiveness [84].

Abstraction development should align with the model's purpose, as explainability is key to the framework. It prioritizes transparency and traceability, highlighting decision-critical aspects of algorithmic outputs. This can be evaluated at the population level, revealing patterns in large datasets, and at the individual level, explaining decisions for specific cases [84].

While "explanation" and "interpretation" are often used interchangeably, a clear distinction exists. Explanation refers to the features used to construct an abstract statement about a decision, while interpretation involves translating that statement into a form understandable by Stakeholders [84]. Within this framework, the interpretability of the system's output is the starting point for evaluation. This is essential for building user understanding, satisfaction, and trust [84].

In clinical settings and, by extension, in system development, the cognitive load imposed by a model's output is only one of many environmental factors. Therefore, outputs must support decision-making without overburdening the user's working memory [84].

Using diagnostic methods to evaluate how helpful a system's output is to the user is essential for guiding system success. Like clinical tools that must avoid overloading healthcare professionals, intelligent systems should deliver information that aligns with user expectations and supports decision-making. A helpful framework must go beyond functionality and account for user experience factors like trust, satisfaction, and comprehension. Trust increases system effectiveness, especially when developers are unsure of its reliability. Understanding how a system works helps users judge when its recommendations are likely right or wrong. Satisfaction reflects how well the system meets expectations. This user-centered, feedback-driven evaluation approach improves complex systems' design, testing, and refinement.

When applying a cognitive model, it is important to ensure the system's output is understandable to both developers and end users. Clear, interpretable results build trust, help users make informed decisions, and allow developers to fine-tune performance. In intelligent and autonomous systems, transparency and justification of decisions are non-negotiable. Without clarity, even accurate results can seem unreliable or misleading [86].

In complex settings like healthcare or autonomous systems development, outputs need to be delivered in ways that match the user's cognitive capacity and decision-making abilities [87]. Does a user's confidence in an autonomous vehicle come from a clear understanding of its reliable performance, or is it merely passive hope that it will function as expected? Ineffective output design that overloads working memory can undermine usability, even if the technology itself is sound [88].

2.6 Evaluating Trust, Satisfaction, and Understanding

Trust, satisfaction, and understanding must be systematically evaluated for a system to be effective and adopted. Trust enhances the system's persuasive power and provides reassurance, particularly when users are uncertain about the system's ability to deliver accurate or reliable outputs [84].

The understanding provided by the model's presentation of the system's perceived reasoning process enables critical evaluation, allowing users and developers to identify when recommendations may be inaccurate. Satisfaction reflects the degree to which the system's output aligns with user expectations and supports their workflow or decision-making process [85].

2.6.1 Building and Measuring Trust

Trust in a system can be strengthened through several key mechanisms. Providing clear explanations of how outputs are generated helps users understand the system's reasoning and increases confidence in its use. Transparency about uncertainty is also important, especially in situations where predictions may be less reliable. Trust further depends on consistent task-specific performance, which should be evaluated through both technical validation and user feedback. These factors should reflect not only the performance metrics established during development but also the user's perceived reliability of the system during real-world operation [85, 86, 88].

2.6.2 Risks of Overtrust

While trust is essential for effective system use, too much trust can lead to cognitive and behavioral risks. Two notable concerns are confirmation bias and automation bias. Confirmation bias occurs when users overlook evidence that contradicts their expectations because the system appears to validate their assumptions. Automation bias arises when users rely too heavily on system recommendations or fail to critically assess the system's output. These risks can be evaluated by analyzing how user decisions align with system recommendations and actual correct outcomes, as well as by measuring the time it takes users to reach conclusions [84, 89, 90].

Determining options and weighing alternatives is fundamental to how humans function and make decisions. This process reflects reasoning as a core cognitive skill. It helps us assess evidence, explore possibilities, and select the most suitable course of action. When systems are designed to incorporate reasoning modeled on human thinking, they must present alternatives and justify them in a way that aligns with how people evaluate choices. An adequate system should support informed decision-making through structured, context-aware reasoning and be capable of evolving with experience. For example, clinical decision support tools in healthcare offer multiple diagnostic paths, each backed by relevant data, enabling physicians to weigh risks, compare treatments, and make judgments tailored to the individual patient [61].

Determining how to enhance the understanding of what a system must do through the abstraction of the reasoning process is essential in systems engineering for developing systems that make informed, context-sensitive decisions. These systems often operate in dynamic, unpredictable environments where fixed rules and predefined responses can pose significant risks.

The modeling of reasoning similar to human decision-making enables systems developers to determine how the system could interpret information within context, adjust to change, and respond appropriately to unexpected conditions. Without this flexibility, system behavior becomes rigid and less dependable, increasing the likelihood of failure in real-world applications [91].

Reasoning about possibilities involves analyzing evidence, constructing arguments, and applying logic to draw conclusions or support decision-making. This process provides an insight pertaining to how practical reasoning supports critical analysis, effective problem-solving, to guide development design decisions. In fields such as medicine, reasoning enables the interpretation of complex data, the evaluation of competing options, and the selection of optimal solutions. These principles apply equally to systems engineering supporting the development of complex systems [92].

In systems engineering, human cognitive processes serve as a model for abstracting design and evaluating the development of complex or intelligent systems. The benefit realized by borrowing from cognition helps in developing an understanding of how people reason, adapt, and make decisions in uncertain and dynamic contexts. Engineers can develop systems with similar capabilities. This approach is critical in domains such as autonomous vehicles, decision-support systems, and adaptive control, where systems must operate reliably under changing conditions and respond appropriately to complex environments [92].

Cognitive modeling supports this effort by guiding test design, requirements analysis, and validation strategies. Given the numerous challenges encountered throughout the engineering process, it is essential to ensure that system development is not only technically sound but also aligned with the principles of human reasoning. This alignment helps create systems that behave in ways that are logical, adaptable, and trustworthy from the user's perspective [93].

By aligning system testing with a diagnostic cognitive model, engineers gain a structured and scalable approach to evaluating intelligent systems; one that mimics human reasoning, prioritizes adaptability, and fosters continuous improvement. This transforms testing from a checklist exercise into a dynamic, feedback-driven process that mirrors how real-world decisions are made and refined [94].

As artificial intelligence and machine learning become increasingly integrated into systems especially those related to diagnostics, decision-making, and autonomous operation, the need to model and simulate higher-order cognitive functions becomes essential. These functions encompass perception, reasoning, learning, adaptation, and decision-making in the face of uncertainty. Cognitive systems seek to mimic or enhance human-like intelligence by incorporating these capabilities into software and hardware, allowing systems to interpret complex data, make informed decisions, and adapt over time which will require further assessment of:

- Trust: Can the system provide consistent, explainable, reliable outputs?
- Ethics: Can the system identify context-sensitive boundaries or flag decisions for human intervention when ethical ambiguity occurs?

These are not simple functions to replicate. Abstracting them requires embedding feedback mechanisms, ensuring transparency, and incorporating human-in-the-loop design principles. This also involves building systems that can adapt in an ethical and socially responsible manner over time, which will likely require further analysis to determine what this entails and what could result in a developed system. However, dedicated metrics that go beyond traditional performance or efficiency measures are required to assess trust. Standard benchmarks often overlook the nuanced, context-sensitive factors that shape trust, such as reliability in uncertain situations, clarity of decision-making, and alignment with human values [95]. These functions are complex to implement. Integrating them into system design demands effective feedback mechanisms, transparency in system behavior, and active involvement of both developers and users in critical decision-making processes.

The complexity and ambiguity become even more apparent as system development evolves and systems must act and adapt ethically and socially responsibly. In this context, does "socially responsible" imply the obligation to design systems that do not cause harm? Ethical behavior in systems may evolve as specific traits of ethics emerge as systems incorporate more human-like cognitive properties [96].

Assessing trust presents unique challenges, requiring dedicated metrics that go beyond conventional measures of performance or efficiency. Standard benchmarks often overlook the subtle, context-dependent factors influencing trust in complex, adaptive systems. Without these specialized measures, evaluating how well a system builds and sustains user confidence in real-world conditions becomes difficult. Trust metrics may include consistency of behavior under stress, transparency of decision-making, clarity of explanations, perceived fairness, user satisfaction over time, and the system's ability to recover from errors, thereby reinforcing user confidence [97].

2.7 Framework: Modeling Cognitive Properties in System Design

Table 2.2 outlines key cognitive properties that can be applied to system design, providing brief descriptions and design examples of how each can be represented or applied. These properties form the foundation for building systems that reflect human behavior, respond appropriately to changing conditions, and support dependable and responsible operations. The framework also supports abstraction and enables a layered evaluation of system components and their relationships, essential for understanding and managing complex, adaptive systems. Consider the following table [86, 95–101]:

Table 2.2 Cognitive properties applied to system design

Cognitive function	Human Equivalent	System abstraction	Testing focus
Perception	Sensory input	Data acquisition (sensors, logs, user input)	Accuracy, fidelity, signal/noise distinction
Pattern recognition	Visual/auditory/semantic awareness	Machine learning, feature extraction	The ability to process and interpret information from visual and auditory sources, as well as to understand the meaning or context behind inputs
Decision-making	Rational analysis	Inference engines, rule-based or probabilistic logic	Validity of outcomes, explainability
Learning	Experience adaptation	Reinforcement learning, continual learning	Rate of knowledge acquisition, sensitivity to shifts in data or behavior, and capacity to adjust to new conditions
Trustworthiness	Reliability, intent recognition	Clear system explanations and reliability indicators	Consistency, transparency, user understanding
Ethical reasoning	Moral and social values	Human-in-the-loop, context rules, constraint modeling	Default fallback behaviors, thresholds for human intervention, and adherence to socially accepted standards

The cognitive properties outlined in the table serve as foundational design considerations for intelligent systems that rely on their own cognitive properties. They do more than describe system capabilities, but can help define how systems should behave under uncertainty, change, and interaction with human users, and would have to consider the following:

- Diagnostics: In healthcare or technical monitoring, systems must integrate reasoning, interpretability, and ethical judgment. This means learning iteratively from data, recognizing uncertainty, and escalating decisions when human oversight is necessary. These capabilities are critical for ensuring that system outputs are accurate, transparent, responsible, and aligned with human expectations [102].
- Adaptation: To reflect the learning and adaptability in human cognition, systems should evolve in how they process data and update their strategies over time. This continuous refinement allows them to respond to changing demands effectively. However, concerns about trust in automation are valid since many technologies in use today

struggle to adapt in transparent and understandable ways to users, which can undermine confidence and hinder adoption [103, 104].

- Trust and Ethics: Trust, ethical judgment, and context awareness are core components of systems that incorporate cognitive elements. These are not secondary concerns; they directly impact how users perceive and rely on a system. Integrating these elements requires the development of appropriate metrics and thorough testing in realistic environments to ensure the system behaves reliably and aligns with both user and developer expectations [105].

Developing reliable systems for real-world applications involves more than meeting technical specifications. The goal is not to replicate human thinking, but to apply key cognitive principles that improve adaptability, clarity, and trust. The future of complex systems depends on thoughtful design that supports intended goals, responds to real-world conditions, and operates responsibly within defined constraints. Abstraction can support system designers in prioritizing specific properties based on the system's intended application, which can help reduce unnecessary complexity; however, does this focused approach bring clarity or risk omitting essential details?

Details that may impact safety and reliability. While finding ways to reduce information overload can help focus the system design, it also raises concerns about whether the full range of behavioral complexity is being captured, which is essential for ensuring that systems are safe and trustworthy [106].

The core challenge in systems engineering is to ensure that systems function reliably while aligning with broader innovation expectations and managing the resulting complexity. Effective governance requires more than high-level design principles. This requires a structured testing protocol and reliable oversight mechanisms that verify whether system behavior remains within acceptable limits under changing and uncertain conditions. These evaluations should go beyond technical performance to include the system's ability to adapt to context-specific requirements, operate transparently, and remain accountable throughout its lifecycle [96].

Achieving a balance between system capability and control involves combining technical methods with structured human feedback, all within clearly defined oversight frameworks. As systems become increasingly complex and exhibit more sophisticated behaviors, continuous governance and management of these behaviors become essential. Human cognitive properties provide a foundational model for abstraction, guiding learning processes and informing system development in a way that supports both performance and accountability [107].

While Asimov's Three Laws of Robotics are often referenced in system design discussions, they are best understood as a conceptual backdrop rather than practical engineering guidance. Asimov used them to explore narrative tensions, not to define technical frameworks. Nonetheless, safety, reliability, and accountability principles remain central to systems engineering. Systems must also operate within a context of trust, particularly as complexity increases. Modeling and abstraction are essential tools for managing this complexity and supporting the development of well-defined methodologies, rulesets, validation processes, and oversight mechanisms to ensure systems perform as required [108].

In systems engineering, managing complexity requires more than just technical knowledge; it necessitates structured thinking, purposeful abstraction, and transparent organization. Discovery and organization are helpful tools that support abstraction by helping engineers identify key components, map relationships, and isolate dependencies. The benefit of a structured approach is that it reduces ambiguity, enabling engineers to focus on what matters most at each stage of system development.

By uncovering relevant elements and organizing them meaningfully, engineers can group related behaviors, filter out unimportant ones, and build simplified models that still capture essential system dynamics. Finding an approach to enhance understanding of system behavior requires refining the developmental analysis and adapting solutions as requirements evolve. Insights from other disciplines, such as medicine, offer valuable parallels. Diagnostic methods routinely handle layers of complexity by integrating genetic, environmental, and cognitive factors. These disciplines demonstrate how interdependent variables can be organized to support accurate and actionable conclusions. In effect, they use discovery and organization to create effective abstractions.

In systems engineering, this approach promotes the use of layered frameworks that help manage complexity, enable informed decision-making under uncertainty, and support system adaptability. Incorporating cross-disciplinary perspectives enhances system design by increasing resilience, improving context awareness, and ensuring more substantial alignment with real-world conditions.

References

1. School of Information, San José State University. (2024). Information organization. https://ischool.sjsu.edu/information-organization
2. Beasley, R. Pyster, A., Stillitto, H., Squires, A., Davidtz, H, Jackson, S., & Wang, Q. (2016). *Organizing business and enterprises to perform systems engineering*. https://www.Sebokwiki.org/wiki/Organizing_Business_andEnterprises_to_Perform_Systems_Engineering
3. Siegenfeld, A. F., & Bar-Yam, Y. (2020). An introduction to complex systems science and its applications. *Complexity, 2020*, Article ID 6105872, 16 pages, Retrieved from online library.wiley.com.
4. Langford, G. (2016). *Engineering systems integration, theory, metrics and methods* (p. 96). CRC Press, Taylor and Francis Group. Retrieved from https://www.researchgate.net/publication/346556661_Engineering_Systems_Integration_Theory_Metrics_and_Methods
5. Blanchard, B., & Boyer, J. (2016). *Systems engineering management* (p. 56, 223). Wiley.
6. Fisher, D. (2006). An emergent perspective on interoperation in systems of systems (Technical Report CMU/SEI-2006-TR-003). Software Engineering Institute.p. 38].
7. Ball, P. (2016). Review of The problems of biological information. *Philosophical Transactions of the Royal Society A Mathematical Physical and Engineering Sciences, 374*(2063), 20150072. Royal Society. Retrieved from: https://www.researchgate.net/publication/293799806_The_problems_of_biological_information

8. Karim, A. S., Brown, D. M., Archuleta, C. M., Grannan, S., Aristilde, L., Goyal, Y., Leonard, J. N., Mangan, N. M., Prindle, A., Rocklin, G. J., Tyo, K. E. J., Zoloth, L., Jewett, M. C., Calkins, S., Kamat, N. P., Tullman-Ercek, D., & Lucks, J. B. (2024). Deconstructing synthetic biology across scales: a conceptual approach for training synthetic biologists. *Nature Communications, 15*(1). Retrieved from: https://www.nature.com/articles/s41467-024-49626-x
9. Zhang, F., Nangreave, J., Liu, Y., & Yan, H. (2014). Structural DNA nanotechnology: State of the art and future perspective [Review of structural DNA nanotechnology: State of the art and future perspective]. *Journal of the American Chemical Society, 136*(32), 11198. American Chemical Society. Retrieved from: https://www.pubs.acs.org/doi/10.1021/ja505101a
10. Garcia, S., & Trinh, C. T. (2019). Review of Modular design: Implementing proven engineering principles in biotechnology. *Biotechnology Advances, 37*(7), 107403. Elsevier. Retrieved from: https://www.sciencedirect.com/science/article/pii/S0734975019300928
11. Pressman, R. S., & Maxim, B. R. (2020). *Software engineering: A practitioner's approach* (9th ed., p. 242). McGraw-Hill Education.
12. Taleb, N. (2012). *Antifragile: Things that gain from disorder* (p. 18). Random House.
13. Santamarina-Campos, V., Montagud, M. Á. C., Esgueva-Lopez, M.-V., & Taroncher-Ballestero, J.-M. (2019, December 16). Evaluation of discovery learning in the field of cultural heritage. In *Proceedings INNODOCT/19. International Conference on Innovation, Documentation and Education.* Retrieved from: https://www.researchgate.net/publication/346007086_Evaluation_of_discovery_learning_in_the_field_of_cultural_heritage
14. Phaal, R., & Müller, G. (2008). An architectural framework for road mapping: Towards visual strategy. *Technological Forecasting and Social Change, 76*(1), 39. Retrieved from: https://www.producao.ufrgs.br/arquivos/disciplinas/187_phaal_2009.pdf
15. Langford, G. O. (2012). *Engineering systems integration: Theory, metrics, and methods* (pp. 182, 221–227, 309–310). https://www.library.oapen.org/bitstream/20.500.12657/40121/1/9781439852880.pdf
16. Hershkowitz, R., Schwarz, B. B., & Dreyfus, T. (2001). Abstraction in context: Epistemic actions. *Journal for Research in Mathematics Education, 32*(2), 195–222. Retrieved from: https://www.researchgate.net/publication/273062252_Abstraction_in_Context_Epistemic_Actions.
17. Blanchard, B., & Blyler, J. (2016). *System engineering management* (5th ed., p. 4, 224, 226, 360). Wiley.
18. Pretorius, P. (2020). Finance functions as a business system. *INCOSE International Symposium, 30*(1), 1606. Retrieved from: https://www.incose.onlinelibrary.wiley.com/doi/10.1002/j.2334-5837.2020.00807.x
19. Osman, N., & Sahraoui, A. (2018). From PLM to ERP: A software systems engineering integration. *International Journal of Software Engineering & Applications, 9*(1), 11. https://doi.org/10.5121/ijsea.2018.9102. Retrieved from: https://www.aircconline.com/ijsea/V9N1/9118ijsea02.pdf
20. Fisher, D. (2006). *CMU/SEI. An emergent perspective on interoperation in systems of systems.* Retrieved from: https://www.kilthub.cmu.edu/articles/report/An_Emergent_Perspective_on_Interoperation_in_Systems_of_Systems/6571907/1?file=12056933
21. Ferris, T. L., Squires, A. F., & Camelia, F. (2015, June). Integrating affective engagement into systems engineering education. Paper presented at 2015 ASEE Annual Conference & Exposition, Seattle, Washington. https://doi.org/10.18260/p.24322. Retrieved from: https://www.peer.asee.org/integrating-affective-engagement-into-systems-engineering-education

22. Hershkowitz, R., Schwarz, B. B., & Dreyfus, T. (2001). Abstraction in context: Epistemic actions. *Journal for Research in Mathematics Education, 32*(2), 195. https://doi.org/10.2307/749673. Retrieved from: https://www-jstor-org.utep.idm.oclc.org/stable/pdf/749673.pdf?refreqid=fastly-default:ae78370cef990610cb8b12a30bc9cfbf&ab_segments=&initiator=&acceptTC=1
23. Dèng, F., Ren, J., & Chen, F. (2018). *Abstraction learning*. arXiv (Cornell University). Retrieved from: https://www.arxiv.org/pdf/1809.03956
24. Santamarina-Campos, V., Montagud, M. Á. C., Esgueva-Lopez, M.-V., & Taroncher-Ballestero, J.-M. (2019, December 16). Evaluation of discovery learning in the field of cultural heritage. *Proceedings INNODOCT/19. International Conference on Innovation, Documentation and Education*. https://doi.org/10.4995/inn2019.2019.10025. Retrieved from: http://www.ocs.editorial.upv.es/index.php/INNODOCT/INN2019/paper/viewFile/10025/5153
25. Wang, Y. (2008). *A hierarchical abstraction model for software engineering*, 43. https://doi.org/10.1145/1370164.1370174. www.researchgate.net/publication/234818080_A_hierarchical_abstraction_model_for_software_engineering
26. Zufferey, M. C. (2018). The abductive art of discovery. *International Journal of Qualitative Methods, Insights From a Qualitative Study on Unaffected Women Managing Genetic Risk of Breast and Ovarian Cancer, 17*(1). https://doi.org/10.1177/1609406917750973. Retrieved from https://www.journals.sagepub.com/doi/epub/10.1177/1609406917750973
27. Shani, A. B., Coghlan, D., & Alexander, B. N. (2019). Rediscovering abductive reasoning in organization development and change research. *The Journal of Applied Behavioral Science, 56*(1), 60. https://doi.org/10.1177/0021886319893016. https://www.d1wqtxts1xzle7.cloudfront.net/107873118/002188631989301620231126-1-8t75tb-libre.pdf?
28. Holyoak, K. J., & Morrison, R. G. (2005). *Cambridge handbook of thinking and reasoning*. Retrieved from https://www.researchgate.net/publication/279852078_The_Cambridge_Handbook_of_Thinking_and_Reasoning
29. An, S., Lin, Z., Chen, B., Fu, Q., Zheng, N., & Lou, J. (2023). *Does deep learning learn to abstract? A systematic probing framework*. arXiv (Cornell University). https://doi.org/10.48550/arxiv.2302.11978 (An et al., 2023) Retrieved from https://www.arxiv.org/abs/2302.11978
30. Bencomo, N., Cabot, J., Chećhik, M., Cheng, B. H. C., Combemale, B., Wąsowski, A., & Zschaler, S. (2024). *Abstraction engineering*. arXiv (Cornell University). https://doi.org/10.48550/arxiv.2408.14074. Retrieved from https://arxiv.org/pdf/2408.14074
31. Wysoki, K, Effective Project Management. 5th. Wiley Publishing, Inc. p468.
32. Wang, Y. (2008). A hierarchical abstraction model for software engineering. In *Proceedings of the 2nd International Workshop on The Role of Abstraction in Software Engineering (ROA '08)*. Association for Computing Machinery, New York, NY, USA, pp. 43–48. Retrieved from: https://www.researchgate.net/publication/234818080_Ahierarchical_abstraction_model_for_software_engineering#fullTextFileContent
33. Hardin, J., & Horton, N. J. (2017). Ensuring That Mathematics is Relevant in a World of Data Science. *Notices of the American Mathematical Society, 64*(9), 986. Retrieved from: https://www.ams.org/journals/notices/201709/rnoti-p986.pdf
34. Wang, Y. (2008). *A hierarchical abstraction model for software engineering*, 43. https://doi.org/10.1145/1370164.1370174. Retrieved from: https://www.researchgate.net/publication/234818080_A_hierarchical_abstraction_model_for_software_engineering#fullTextFileContent
35. Daniluk, A. (2011). Visual modeling for scientific software architecture design. A practical approach. *Computer Physics Communications, 183*(2), 213. https://doi.org/10.1016/j.cpc.2011.07.021]

36. Doe, R. (2020). An open, integrated modular format: For flexible and intelligible architecture, engineering and construction design and production. *International Journal of Architectural Computing, 19*(1). Retrieved from: https://www.researchgate.net/publication/343217201_An_open_integrated_modular_format_For_flexible_and_intelligible_architecture_engineering_and_construction_design_and_production
37. Langford, G. O. (2012). *Engineering systems integration: Theory, metrics, and methods* (p. 138). Retrieved from https://www.researchgate.net/publication/346556661_Engineering_Systems_Integration_Theory_Metrics_and_Methods.
38. Holyoak, K. J., & Morrison, R. G. (2005). *Cambridge handbook of thinking and reasoning* (p. 36). Cambridge University Press
39. Ranaldi, L., & Pucci, G. (2023). Knowing knowledge: Epistemological study of knowledge in transformers. *Applied Sciences*. Retrieved from: https://www.mdpi.com/2076-3417/13/2/677
40. Saad, W., Hashash, O., Thomas, C. K., Chaccour, C., Debbah, M., Mandayam, N. B., & Han, Z. (2024). *Artificial general intelligence (AGI)-native wireless systems: A journey beyond 6G*. arXiv (Cornell University). Retrieved from: https://arxiv.org/abs/2405.02336
41. Moore, J., Tucci, L., & Techtarget. (2024). *What is the metaverse? An explanation and in-depth guide*. Retrieved from: https://www.cdn.ttgtmedia.com/rms/editorial/Metaverse.pdf
42. Saad, W., Hashash, O., Thomas, C. K., Chaccour, C., Debbah, M., Mandayam, N. B., & Han, Z. (2024). Artificial general intelligence (AGI)-native wireless systems: A journey beyond 6G. arXiv (Cornell University). Retrieved from:https://arxiv.org/abs/2405.02336].
43. Krawczyk, D. C. (2018). Chapter 10—analogical reasoning. In D. C. Krawczyk (Ed.), *Reasoning* (pp. 227–253). Academic Press. ISBN 9780128092859, Retrieved from https://www.sciencedirect.com/science/article/abs/pii/B9780128092859000107
44. Aboelela, S. W., Larson, E., Bakken, S., Carrasquillo, O., Formicola, A. J., Glied, S., Haas, J., & Gebbie, K. M. (2006). Defining interdisciplinary research: Conclusions from a critical review of the literature [Review of defining interdisciplinary research: Conclusions from a critical review of the literature]. *Health Services Research, 42*, 329. Retrieved from https://www.onlinelibrary.wiley.com/doi/10.1111/j.1475-6773.2006.00621.x
45. Farid, T., Ali, S., Sajid, M., & Akhtar, K. P. (2021). Sustainability of project-based learning by incorporating transdisciplinary design in fabrication of hydraulic robot arm. *Sustainability, 13*(14), 7949. Retrieved from: https://www.onlinelibrary.wiley.com/doi/10.1111/j.1475-6773.2006.00621.x
46. Georgeff, L., & Bessiere, P. (1985). *A procedural logic*. Artificial Intelligence Center, SRI International. Retrieved from: https://www.ijcai.org/Proceedings/85-1/Papers/099.pdf
47. Automated Driving Systems: A Vision for Safety. 2.0. (2017). National Highway Traffic Safety Administration (NHTSA). Retrieved from https://www.nhtsa.gov/sites/nhtsa.gov/files/documents/13069a-ads2.0_090617_v9a_tag.pdf
48. Kalra, N., & Paddock, S. M. (2016). *Driving to safety: How many miles of driving would it take to demonstrate autonomous vehicle reliability*. RAND Corporation. Retrieved from: https://www.rand.org/pubs/research_reports/RR1478.html
49. SAE International. (2021). *Taxonomy and definitions for terms related to driving automation systems for on-road motor vehicles* (J3016_2021). Retrieved from: https://www.sae.org/standards/content/j3016_202104
50. National Institute of Standards and Technology (NIST). (2016). *Autonomy levels for unmanned systems (ALFUS) framework*. U.S. Department of Commerce. Retrieved from: https://www.nist.gov/system/files/documents/el/isd/ks/NISTSP_1011-I-2-0.pdf
51. National Highway Traffic Safety Administration (NHTSA). (2020). *Automated vehicles for safety*. https://www.nhtsa.gov/technology-innovation/automated-vehicles.

52. Thrun, S., Montemerlo, M., Dahlkamp, H., et al. (2006). Stanley: The robot that won the DARPA grand challenge. *Journal of Field Robotics, 23*(9), 661–692. Retrieved from: https://www.onlinelibrary.wiley.com/doi/pdf/10.1002/rob.20147
53. Petit, J., & Shladover, S. E. (2015). Potential cyberattacks on automated vehicles. *IEEE Transactions on Intelligent Transportation Systems, 16*(2), 546–556. Retrieved from: https://www.ieeexplore.ieee.org/document/6899663
54. Paden, B., Čáp, M., Yong, S. Z., Yershov, D., & Frazzoli, E. (2016). A survey of motion planning and control techniques for self-driving urban vehicles. *IEEE Transactions on Intelligent Vehicles, 1*(1), 33–55. Retrieved from: https://www.ieeexplore.ieee.org/stamp/stamp.jsp?tp=&arnumber=7490340
55. Bagschik, G., Menzel, T., & Maurer, M. (2018). Ontology based scene creation for the development of automated vehicles. In *2018 IEEE Intelligent Vehicles Symposium (IV)* (pp. 1813–1820). Retrieved from: https://www.ieeexplore.ieee.org/stamp/stamp.jsp?tp=&arnumber=8500501
56. Koopman, P., & Wagner, M. (2016). Challenges in autonomous vehicle testing and validation. *SAE International Journal of Transportation Safety, 4*(1), 15–24. Retrieved from: https://www.researchgate.net/publication/301272911_Challenges_in_Autonomous_Vehicle_Testing_and_Validation
57. Vahdat-Nejad, H., Ramazani, A., Mohammadi, T., & Mansoor, W. (2016). A survey on context-aware vehicular network applications. *Vehicular Communications, 3*, pp. 43–57. ISSN 2214-2096. Retrieved from: https://www.sciencedirect.com/science/article/abs/pii/S2214209616000036
58. NIST. (2024). NIST cybersecurity framework (CSF) 2.0, NIST CSWP 29*. Retrieved from https://www.nvlpubs.nist.gov/nistpubs/CSWP/NIST.CSWP.29.pdf
59. Meadows, D. H. (2009). Earthscan. *Thinking in systems* (p. 35).
60. Nakagawa, T. (2021). *From the four-box scheme of abstraction in science and technology to the six-box scheme for creative problem solving.* Retrieved from: https://www.linkedin.com/pulse/from-four-box-scheme-abstraction-science-technology-six-box-nakagawa/
61. Terreros, D. A. (2015). *Conversations*. Texas Tech University Health Sciences Center.
62. Hanane, T., Kristin, M., Philip, L., De, J. (2023). Differential impact of environmental factors on systemic and localized autoimmunity. *Frontiers in Immunology.* Retrieved from: https://www.frontiersin.org/journals/immunology/articles/10.3389/fimmu.2023.1147447/full
63. Hood, L., Heath, J. R., Phelps, M. E., & Lin, B. (2004). Systems biology and new technologies enable predictive and preventative medicine. *Science, 306*(5696), 640–643. Retrieved from https://www.sccs.swarthmore.edu/users/08/bblonder/phys120/docs/hood.pdf
64. Kitano, H. (2002). Systems biology: A brief overview. *Science, 295*(5560), 1662–1664. Retrieved from https://www.researchgate.net/publication/11489394_Systems_Biology_A_Brief_Overview
65. Blanchard, B. S., & Fabrycky, W. J. (2010). *Systems engineering and analysis* (5th ed., pp. 19, 41). Pearson.
66. Boehm, B., & Lane, J. A. (2007). Using the incremental commitment model to integrate system acquisition, systems engineering, and software engineering. *CrossTalk: The Journal of Defense Software Engineering.* Retrieved from: https://www.researchgate.net/publication/228699789_Using_the_incremental_commitment_model_to_integrate_system_acquisition_systems_engineering_and_software_engineering
67. Maier, M. W., & Rechtin, E. (2009). *The art of systems architecting* (3rd ed., pp. 218–219). CRC Press.

68. Frumkin, H. (2005). Environmental health: From global to local. Jossey-Bass. Retrieved from: https://www.google.com/books/edition/Environmental_Health/ASzvCQAAQBAJ?hl=en&gbpv=1&dq=Frumkin,+H.+(2005).+Environmental+health:+From+global+to+local.+Jossey-Bass.&pg=PR9&printsec=frontcover
69. Rappaport, T. S. (2014). *Wireless communications: Principles and practice* (2nd ed.). Prentice Hall. Retrieved from: https://www.google.com/books/edition/Wireless_Communications/CRrxEAAAQBAJ?hl=en&gbpv=1&dq=Rappaport,+T.+S.+(2014).+Wireless+communications:+Principles+and+practice+(2nd+ed.).+Prentice+Hall.&pg=PA1&printsec=frontcover
70. MIL-STD-810H. (2019, January 31). Retrieved from https://www.cvgstrategy.com/wp-content/uploads/2019/03/MIL-STD-810H.pdf, pp4,871-912,347,801.
71. Janeway, C. A., Travers, P., Walport, M., & Shlomchik, M. J. (2001). *Immunobiology: The immune system in health and disease* (5th ed., p. 14). Garland Science.
72. Hofmeyr, S. A., & Forrest, S. (2000). Architecture for an artificial immune system. *Evolutionary Computation, 8*(4), 443–473. Retrieved from https://www.researchgate.net/publication/12197865_Architecture_for_an_Artificial_Immune_System
73. Sommer, R., & Paxson, V. (2010). Outside the closed world: On using machine learning for network intrusion detection. *IEEE Symposium on Security and Privacy*, 305–316. Retrieved from https://www.ieeexplore.ieee.org/stamp/stamp.jsp?tp=&arnumber=5504793
74. Endy, D. (2005). Foundations for engineering biology. *Nature, 438*(7067), 449–453. https://doi.org/10.1038/nature04342. Retrieved from: http://www.icampus.mit.edu/files/2011/11/iGem-Nature-Review-Foundns.pdf
75. INCOSE. (2015). *Systems engineering handbook: A guide for system life cycle processes and activities* (4th ed.). Wiley. Retrieved from: https://www.download.e-bookshelf.de/download/0003/6422/62/L-G-0003642262-0007580512.pdf
76. NIST SP 800-30 Rev. 1 (2012). *Guide for conducting risk assessments.* U.S. National Institute of Standards and Technology. Retrieved from: https://www.nvlpubs.nist.gov/nistpubs/Legacy/SP/nistspecialpublication800-30r1.pdf
77. Kankam, H. (2025). A brief excursion into human cognition, the evolving influence of social media & artificial. *Intelligence.* https://doi.org/10.1007/978-3-031-89752-8
78. Harada, T., Shimizu, T., Kaji, Y., Suyama, Y., Matsumoto, T., Kosaka, C., Shimizu, H., Nei, T., & Watanuki, S. (2020). A perspective from a case conference on comparing the diagnostic process: Human diagnostic thinking vs. artificial intelligence (AI) decision support tools. *International Journal of Environmental Research and Public Health, 17*(17), 6110. Retrieved from: https://www.pmc.ncbi.nlm.nih.gov/articles/PMC10916499/#:~:text=AI can assist healthcare professionals, satisfaction and involvement of users.
79. Nguyen, X. V., Öztek, M. A., Nelakurti, D. D., Brunnquell, C. L., Mossa-Basha, M., Haynor, D. R., & Prevedello, L. M. (2020). Applying artificial intelligence to mitigate effects of patient motion or other complicating factors on image quality, review of applying artificial intelligence to mitigate effects of patient motion or other complicating factors on image quality. *Topics in Magnetic Resonance Imaging, 29*(4), 175. Lippincott Williams & Wilkins. Retrieved from: https://www.journals.lww.com/topicsinmri/fulltext/2020/08000applying_artificial_intelligence_to_mitigate.3.aspx
80. Hubert, K., Awa, K. N., & Zabelina, D. L. (2024). The current state of artificial intelligence generative language models is more creative than humans on divergent thinking tasks. *Scientific Reports, 14*(1). Retrieved from: https://www.nature.com/articles/s41598-024-53303-w
81. Shen, J., Zhang, C. J. P., Jiang, B., Chen, J., Song, J., Liu, Z., He, Z., Wong, S. Y., Fang, P.-H., & Ming, W. (2019). Artificial intelligence versus clinicians in disease diagnosis: Systematic review. *JMIR Medical Informatics, 7*(3). JMIR Publications. Retrieved from: https://www.medinform.jmir.org/2019/3/e10010/

82. Fenwick, A., Molnár, G., & Frangos, P. (2024). Revisiting the role of HR in the age of AI: bringing humans and machines closer together in the workplace [Review of Revisiting the role of HR in the age of AI: bringing humans and machines closer together in the workplace]. *Frontiers in Artificial Intelligence, 6*. Frontiers Media. Retrieved from: http://www.frontiersin.org/journals/artificial-intelligence/articles/10.3389/frai.2023.1272823/full
83. Artificial intelligence and machine learning in armed conflict: A human-centred approach. (2020). *International Review of the Red Cross, 102*(913), 463. Retrieved from: https://www.icrc.org/en/document/artificial-intelligence-and-machine-learning-armed-conflict-human-centred-approach
84. Wysocki, O., Davies, J. K., Vigo, M., Armstrong, A., Landers, D., Lee, R., & Freitas, A. (2022). Assessing the communication gap between AI models and healthcare professionals: Explainability, utility and trust in AI-driven clinical decision-making. *Artificial Intelligence, 316*, 103839. Retrieved from: https://www.sciencedirect.com/science/article/pii/S0004370222001795?via%3Dihub
85. Amershi, S., Chickering, M., Drucker, S. M., Lee, B., Simard, P., & Suh, J. (2019). Guidelines for human-AI interaction. In *Proceedings of the 2019 CHI Conference on Human Factors in Computing Systems*. Retrieved from: https://www.researchgate.net/publication/332742200_Guidelines_for_Human-AI_Interaction.
86. Doshi-Velez, F., & Kim, B. (2017). *Towards a rigorous science of interpretable machine learning*. Retrieved from https://arxiv.org/pdf/1702.0860
87. Wickens, C. D., Hollands, J. G., Banbury, S., & Parasuraman, R. (2013). *Engineering psychology and human performance* (4th ed., p. 247). Psychology Press. Retrieved from: https://www.scribd.com/document/501890445/Engineering-Psychology-and-Human-Performance
88. Ghassemi, M., Oakden-Rayner, L., & Beam, A. L. (2020). The false hope of current approaches to explainable artificial intelligence in health care. *The Lancet Digital Health, 3*(11), e745–e750. Retrieved from: https://www.thelancet.com/journals/landig/article/PIIS2589-7500
89. Klein, G. (2008). Naturalistic decision making. *Human Factors, 50*(3), 456–460. Retrieved from: https://www.researchgate.net/publication/319770380_Naturalistic_decision_making
90. Wickens, C. D., Hollands, J. G., Banbury, S., & Parasuraman, R. (2015). *Engineering psychology and human performance* (4th ed., p. 543). Psychology Press. Retrieved from: https://archive.org/details/engineeringpsych0000wick/mode/2up?q=automation+bias
91. What Is Logic In Programming | Restackio. (2024). Retrieved from: https://www.restack.io/p/combinatorial-applications-in-software-engineering-answer-logic-in-programming
92. Ho, S., Kim, S., & Kim, S. (2025). Uncertainty-aware multimodal trajectory prediction via a single inference from a single model. *Sensors, 25*(1), 217. Retrieved from https://www.mdpi.com/1424-8220/25/1/217
93. Siemens, G., Marmolejo-Ramos, F., Gabriel, F., Medeiros, K. E., Marrone, R., Joksimovíc, S., & Laat, M. de. (2022). Human and artificial cognition. *Computers and Education Artificial Intelligence, 3*, 100107. Retrieved from: https://www.sciencedirect.com/science/article/pii/S2666920X22000625?via%3Dihub
94. Ke, Z., Jiao, F., Ming, Y., Nguyen, X.-P., Xu, A., Long, D. X., Li, M., Qin, C., Wang, P., Savarese, S., Xiong, C., & Joty, S. (2025). *A survey of frontiers in LLM reasoning: Inference scaling, learning to reason, and agentic systems*. Retrieved from: https://arxiv.org/pdf/2504.09037
95. Gunning, D., & Aha, D. (2019). DARPA's explainable artificial intelligence (XAI) program. *AI Magazine, 40*(2), 44–58. Retrieved from https://onlinelibrary.wiley.com/doi/epdf/10.1609/aimag.v40i2.2850 in complex, adaptive systems.

96. Shneiderman, B. (2020). Bridging the gap between ethics and practice: Guidelines for reliable, safe, and trustworthy human-centered AI systems. *ACM Transactions on Interactive Intelligent Systems, 10*(4), 1–31. Retrieved from: https://dl.acm.org/doi/10.1145/3419764. https://www.researchgate.net/publication/347129395_Bridging_the_Gap_Between_Ethics_and_Practice_Guidelines_for_Reliable_Safe_and_Trustworthy_Human-centered_AI_Systems
97. Hoff, K. A., & Bashir, M. (2015). Trust in automation: Integrating empirical evidence on factors that influence trust. *Human Factors, 57*(3), 407–434. Retrieved from: https://www.researchgate.net/publication/272887576_Trust_in_Automation_Integrating_Empirical_Evidence_on_Factors_That_Influence_Trust. https://doi.org/10.1177/0018720814547570
98. Anderson, M., & Anderson, S. L. (2011). Machine ethics: Creating an ethical intelligent agent. *AI Magazine, 28*(4), 15–26. Retrieved from: https://ojs.aaai.org/aimagazine/index.php/aimagazine/article/view/2065
99. Endsley, M. R. (1995). Toward a theory of situation awareness in dynamic systems. *Human Factors, 37*(1), 32–64. Retrieved from: https://www.researchgate.net/publication/210198492_Endsley_MR_Toward_a_Theory_of_Situation_Awareness_in_Dynamic_Systems_Human_Factors_Journal_371_32-64
100. Lee, J. D., & See, K. A. (2004). Trust in automation: Designing for appropriate reliance. *Human Factors, 46*(1), 50–80. Retrieved from: https://user.engineering.uiowa.edu/~csl/publications/pdf/leesee04.pdf
101. Russell, S., & Norvig, P. (2021). *Artificial intelligence: A modern approach* (4th ed., pp. 840–853). Pearson.
102. Leveson, N. (2012). *Engineering a safer world: Systems thinking applied to safety*. MIT Press.
103. Retrieved from: https://www.researchgate.net/publication/329654446_Engineering_a_Safer_World_Systems_Thinking_Applied_to_Safety
104. Hoffman, R. R., Johnson, M., Bradshaw, J. M., & Underbrink, A. (2013). Trust in automation. *IEEE Intelligent Systems, 28*(1), 84–88. Retrieved from: https://www.researchgate.net/publication/260305028_Trust_in_Automation
105. Textor, C., Zhang, R., Lopez, R., Schelble, B.G., McNeese, N. J., Freeman, G., Pak, R., Tossell, C., & de Visser, E. J. Exploring the relationship between ethics and trust in human–artificial intelligence teaming: A mixed methods approach.
106. Simon, H. A. (1996). *The sciences of the artificial* (3rd ed., p. 23). MIT Press. Retrieved from: https://journals.sagepub.com/doi/epub/10.1177/15553434221113964
107. Christiano, P. F., Leike, J., Brown, T. B., Martic, M., Legg, S., & Amodei, D. (2017). Deep reinforcement learning from human preferences. In *Advances in neural information processing systems* (pp. 4299–4307). Retrieved from: https://arxiv.org/pdf/1706.0374
108. Anderson, S. L. (2005). *Asimov's "Three Laws of Robotics" and machine metaethics*. University of Connecticut Dept. of Philosophy. Retrieved from: https://cdn.aaai.org/Symposia/Fall/2005/FS-05-06/FS05-06-002.pdf

3 Causality and Composability

Causality and composability are essential concepts in systems engineering, offering a foundation for analyzing the complex relationships and dynamic behaviors that arise during system development. Causality, in particular, captures the cause-and-effect dependencies between components and processes, helping engineers understand how changes in one part of the system can influence others [1, 2].

Composability, on the other hand, addresses the capacity to construct intricate systems by seamlessly integrating modular and reusable components, ensuring that the emergent behavior of the composite system remains predictable and aligned with design objectives. The importance of causality in engineering systems cannot be overstated, as it provides a basis for predicting performance, identifying potential failure modes, and improving overall system reliability [3].

One major attribute of composability is the ability to combine and interconnect individual components or modules to create a larger, more complex system. Essentially, everything is interconnected, as evident by our daily interactions with networks and cloud services that provide services as needed. Composability is about designing systems where components can be easily assembled, reconfigured, and reused to meet different needs [4].

Causality and composability, in the context of abstraction and hierarchy, help manage complexity by structuring interactions and dependencies across different layers. Through abstraction, essential interactions derive from causal relationships, facilitating clearer insights into how component behaviors affect system-wide outcomes. Hierarchical design, meanwhile, leverages composability to integrate modular components predictably, ensuring that causal effects are traceable and manageable across multiple levels of system

F. A. Macias, *Project Development Simplified*, Synthesis Lectures on Engineering, Science, and Technology, https://doi.org/10.1007/978-3-032-01043-8_3

architecture. This synergy enables robust system design, precise failure analysis, and scalable development, reinforcing reliability and alignment with engineering objectives.

In systems engineering, when each function is executed correctly by its respective subsystem and external conditions are controlled, the outcome becomes predictable and reliable. This illustrates how causality and composability interact, where small, well-defined actions combine to produce the final system behavior.

3.1 Causality

Causality refers to the relationship between different components and interactions within a system that lead to specific outcomes or behaviors. Understanding causality in engineering systems involves identifying the cause-and-effect relationships between various elements and events to predict and control system behavior.

A formal framework for understanding causality that includes causal diagrams and equations describing how changes in one variable propagate through a system to affect others. The relational way to represent causality is through the following equation that links a cause C with its effect E as follows [5]:

$$E = f(C, X)$$

where:

E The effect or outcome observed.
C The primary cause or influencing factor.
X A set of external variables or conditions that might modulate the effect.
f A function that defines the nature of the relationship.

For example, if the temperature setting on a thermostat (X) is turned up, increasing the temperature, then the room should get warmer (Y), depending on things like current temperature (Z), which in this case results in:

$$Y = f(X, Z)$$

This indicates that Y (the effect) is determined by X (the cause) and may also be influenced by Z (extra conditions), offering a straightforward perspective on cause and effect in systems.

Causality in systems engineering emphasizes the relationships between system components and their interactions, helping shape specific outcomes and behaviors. Understanding causality within engineering systems requires identifying cause-and-effect relationships among various elements and processes, which supports the potential to predict and manage system behavior when necessary [6].

A comprehensive understanding of causality aids in systems development by enabling engineers to pinpoint critical interactions among system components through the analysis of cause-and-effect relationships, facilitating informed prioritization of design efforts and resource allocation to the areas with the most substantial impact on system performance. By understanding causality, engineers can foresee how changes in one part of the system might ripple through and affect the overall system behavior, allowing them to anticipate potential issues, fine-tune performance, and bolster system reliability through simulations and consideration of causal relationships [6].

Processes, subsystems, or other systems can influence the overall functionality and are important for understanding causality because they reveal how events are logically linked, supporting the analysis to help identify cause and effect. The order in which processes within a system occur establishes causal relationships, distinguishing them from simple correlations.

While correlation indicates a fundamental connection, causality is characterized by a structured sequence in which one event directly influences another. These chains of influence create clear pathways that should be planned or may emerge based on system functionality and dependencies [6].

Several types of causality are relevant to systems engineering and can be categorized as follows [6]:

- Proximate Causality refers to immediate, localized causes that directly influence system behavior. These causes are often identifiable through direct observation or analysis and are typically the focus of troubleshooting or failure analysis.
- Conditional Causality addresses causes that depend on specific conditions or contexts. They are not always immediately apparent and may require a deeper understanding of the system's operational context or planning data for identification, and often involve complex interactions within the system.
- Objective Causality encompasses causes that are based on measurable and observable circumstances and tends to rely on empirical data and quantitative analyses to help establish cause-effect relationships within systems.

In addition, objective causalities represent interactions among known system components, functions, and their relationships. Tracing these interactions step by step can be challenging due to the complexity and scale of system development. However, to effectively analyze these causal linkages, engineers may need to employ systematic approaches, such as event correlation, fault tree analysis, and system dynamics modeling, to identify behaviors that signify causal relationships. Causality is the relationship between two events, where one event uniquely determines the next event [6].

For system development, the functionality of objective causalities is tightly linked to governance structures, policies, and regulatory frameworks that define system behavior throughout its lifecycle. These causalities enable mapping management decisions and policy implementations to specific system requirements, facilitating impact assessments, failure analysis, and lifecycle sustainment strategies. In addition, objective causalities support structured analyses of process-driven events, enabling engineers to identify critical decision points that affect overall system performance, reliability, and compliance with regulatory requirements [6].

Explicit causal assumptions help eliminate ambiguities that often arise from methods solely based on statistical correlations or in measuring the strength and direction of the relationship between two variables, indicating how changes in one may affect another [7].

In contrast to purely correlational analysis, which identifies relationships without indicating the direction of influence, explicit causal models enable engineers to identify the direct effects of design decisions, process modifications, and interactions within systems. This insight is invaluable for making informed choices in complex engineered systems, as unnoticed dependencies can lead to failure if not adequately identified. The benefit of incorporating causal mechanisms within a system development is that the outcome could be more effective than relying on other measures of causality and structural limitations. This makes it easier to understand system relationships and patterns of behavior, and to make accurate predictions of performance [8].

In systems engineering practice, this means leveraging models such as causal loop diagrams, dependency matrices, and system dynamics models to explicitly map out cause-and-effect pathways. These models facilitate the identification of critical leverage points, understanding of failure propagation, and optimization of control strategies to prevent undesired outcomes. Understanding these disruptions enables engineers to predict system responses and effectively re-stabilize operations. In essence, explicit causal modeling in systems engineering not only enhances reliability and clarity but also equips engineers to handle complex, interconnected systems with higher confidence and predictive capability.

Confidence and predictability infer the use of causal artificial intelligence methods to use high-level quantitative models to represent networks of causal mechanisms, enabling the prediction of intervention outcomes within complex systems [9].

These approaches rely on causal inference to establish direct cause-and-effect relationships between interventions and their outcomes, guided by structured data analysis. This process incorporates a range of estimation techniques, systematically applied across the design, implementation, and analysis phases to ensure precision and reliability.

In the context of systems engineering, causal inference supports effective decision-making by mapping dependencies and understanding how changes to one subsystem may impact overall system performance.

Complementing this, causal discovery is utilized to comprehend and model causality across interconnected system variables. Using causal graphs, engineers can visualize dependencies, feedback loops, and points of vulnerability within a system, aiding in both optimization and risk assessment [10].

This method decomposes an intervention's total impact into its causal and non-causal components, thereby reducing estimation bias and enhancing predictive accuracy. Such clarity enables system engineers to fine-tune designs and adapt to evolving conditions without compromising performance. Additionally, causal modeling serves as a critical tool for validating system behavior against theoretical predictions. By hypothesizing specific interactions and testing them against empirical data, engineers can confirm the integrity and reliability of system components and their interactions.

This approach is essential for verifying that the observed outputs align with design expectations and project specifications. While data-driven methods help minimize statistical bias, assessing discrepancies between estimated and actual system behavior often requires advanced techniques beyond traditional estimation methods [11].

Understanding causality is helpful for improving system design, predicting performance, identifying failure modes, and enhancing overall system reliability Analyzing cause-and-effect relationships within a system helps identify critical interactions between system components, which assists in prioritizing design task and allocating resources to functions with the most significant impact on system performance [12].

To effectively design and analyze complex systems, engineers should prioritize understanding causality. This insight allows them to predict how changes in one part of the system may impact overall behavior, anticipate potential issues, and optimize performance. Incorporating causal analysis enables more reliable simulations, supporting engineers in evaluating different scenarios and identifying critical vulnerabilities before they lead to failures. Emphasizing causality in the design process not only strengthens system reliability but also enhances safety and resilience [13].

To minimize the risk of failure, system-level testing becomes critical, even though it is often challenging due to the complex interactions among components, each exhibiting its own spatial and temporal behaviors that must occur to maintain necessary functionality and performance. Effective testing requires an understanding of these interdependencies and anticipation of how changes in one area may influence the entire system. Systemic design builds upon the principles of complex systems, recognizing characteristics such as self-organization, feedback loops, and emergence.

Adopting this approach emphasizes the interconnectedness and dynamic nature of system components, encouraging engineers and designers to view individual elements as part of a larger, interdependent whole. Additionally, it acknowledges that problems often have multiple causes and definitions, requiring holistic analysis and adaptive solutions. By prioritizing systemic design, more resilient, responsive, and sustainable engineering solutions are created [14].

Complexity demands collaborative efforts to build an understanding of the system and remain open to reframing problems as new insights emerge [15]. The concept of causality is fundamental to understanding how engineering systems function, evolve, and respond to both internal and external influences [16].

By understanding the causal relationships within a system, engineers can enhance system performance, anticipate potential failures, and improve overall reliability [6]. Causality is a versatile concept that spans various scientific disciplines, including physics, medicine, logic, philosophy, law, psychology, and linguistics [17]. In the context of system dynamics, causality becomes especially valuable, as it provides clear explanations for how and why systems behave the way they do, enabling more effective modeling and optimization [6].

While root cause analysis is a valuable tool for identifying the origins of system issues, it has potential drawbacks. One major concern is its tendency to overemphasize specific factors, potentially overlooking broader relationships and interactions within different layers of system design [6].

Traditional statistical debugging methods, which rely on correlational predicates, are often employed to diagnose system faults. However, these techniques face limitations due to their dependence on large-scale datasets and the challenges associated with recognizing patterns in high-dimensional spaces [11]. As a result, purely statistical approaches may struggle to capture the deeper causal mechanisms driving system behavior.

Additionally, when interaction serves as the cement of systems, the causality of events emerges from the interplay of objects, mechanisms, and behaviors that converge or coincide. It is important to recognize that although causality drives events, it is not necessarily the origin of causality itself. The relationships between causality and mechanisms are often obscured and complicated by confounding symptoms of causal factors [6].

Causality and correlation represent two distinct concepts in system analysis. Causality is characterized by a sequence of events where one event directly influences the occurrence of another, establishing a cause-and-effect relationship [6]. In contrast, correlation merely indicates a connection or pattern between events without proving that one event causes the other [6].

Causality depends on the notion that specific conditions must be met for one event to trigger another, clearly defining the requirements for necessary and sufficient events. Correlated events, however, do not demonstrate this level of specificity, as their association does not confirm direct influence [6]. The fundamental purpose of causality is to outline a predictable sequence of events where one occurrence reliably leads to the next, offering clarity and direction in system behavior [6].

3.2 Composability

Composability is the capability of system components to be assembled, integrated, and reconfigured in multiple ways to generate new functionalities. This concept is linked to causality, where modifications to one part of the system can result in changes in others. For example, in manufacturing, altering input materials or adjusting process parameters can lead to measurable differences in the final product, highlighting the causal dependencies within the system. The modular nature of composable systems allows these causal effects to function efficiently across different layers of the system architecture, enhancing adaptability and scalability in complex engineered systems. This interconnected framework supports real-time adjustments and optimizes system performance by enabling exact control over individual components [18].

The hierarchical nature of systems and the ability to abstract away details at different levels enable engineers to manage the complexity of large-scale systems by focusing on the interactions between components rather than the intricate details of each individual element. The principles of causality and composability, when applied together, support the development of abstraction and hierarchy in systems engineering, allowing for more effective design, analysis, and optimization of complex systems [19].

Leveraging existing sub-system models to compose a new full system model significantly reduces the time and domain expertise required to model a complex interaction system, increasing reusability and extensibility. In essence, composability allows for the creation of complex systems by combining simpler, well-defined components, making it easier to manage and understand the overall system behavior.

Loose coupling between component models facilitates parallel model development, reducing the time required to build a full system model, and enabling easier reuse and alteration of models. This is particularly important when incorporating new elements or adapting to changing requirements, as changes to one component have relatively less effect on other components [1]. Component-based design revolves around two general principles [20]:

- Interface-based integration: When a system developer uses a component for an application, there is a reliance on what is guaranteed by the interface, allowing another implementation to be substituted if needed without modifying the rest of the application.
- Reusability: Once a component is implemented, a system developer can reuse the component in multiple applications without changing the component itself.

Loose coupling in component models allows parallel model development, reducing the time required to build a full system model and enabling easier reuse and alteration of models [6]. In the context of complex systems, composability helps manage the interactions and dependencies that arise from the system's multi-scale nature [21]. Engineers can more effectively analyze, design, and optimize the system's behavior by dividing a complex system into loosely interdependent component models.

Composability enhances flexibility by enabling systems to be reconfigured or adapted to meet evolving requirements or conditions. Well-defined interfaces and protocols simplify complexity by ensuring predictable interactions between components. Additionally, composability streamlines maintenance and upgrades, allowing individual components to be replaced or updated without affecting the overall system. It enables the selection and combination of components in different configurations to meet user requirements [6].

3.3 Composability and Systems Engineering

Composability builds upon the principles of modularity and standardization to enable the integration of independently developed components into a unified system [6]. This approach promotes flexibility by allowing systems to be reconfigured or adapted to meet evolving requirements or conditions. Composability reduces complexity through well-defined interfaces and protocols, allowing components to interact predictably. By streamlining planning and development, individual system components can be introduced or updated without disrupting the entire system development. At its core, composability represents the capability to effectively select and combine components in different configurations to meet user requirements.

3.3.1 Benefits of Composability

Leveraging existing subsystem models to compose a new complete system model significantly reduces the time and domain expertise required to model complex interaction systems. Composability enables the creation of complex systems by combining simpler, well-defined components, making system behavior easier to manage and understand. Loose coupling between component models facilitates parallel development, reducing the time required to build a complete system model and enabling easier reuse and adaptation of models. This is especially valuable when integrating new elements or adjusting to changing requirements, as modifications to one component have minimal impact on others [1].

3.3.2 Principles of Component-Based Design

Component-based design is built on two core principles: interface-based integration and reusability. Interface-based integration ensures that when developers use a component, they depend only on the guarantees provided by its interface. This allows different implementations to be swapped in without requiring changes to the rest of the system [20].

Reusability means that once a system, subsystem, or component is developed, it can be applied across multiple applications without modification. Supporting these principles is the concept of loose coupling, which aims to minimize dependencies between components. In a loosely coupled system, components communicate through clearly defined interfaces without needing to understand each other's internal workings. This separation simplifies planning, development, testing, modification, and replacement, helping to maintain the overall stability of the system [20].

This approach promotes flexibility and parallel development, simplifying upgrades and enabling seamless component replacements as requirements change [6].

3.4 Composability and Modularity in System Design

Composability is a key principle in system design that enables the creation of complex systems by integrating simpler, self-contained components [6]. This approach allows systems to adapt to changing requirements, accommodate new technologies, and scale effectively. Composable modeling decomposes large-scale system models into loosely connected component models, similar to modular software design, which reduces the impact of changes in one component on others [6]. Leveraging existing subsystem models to create new system models minimizes the time and expertise required for complex interactions [6].

Modularity supports composability by structuring systems into distinct, well-defined modules with a standardized interface [22]. Loose coupling between these modules allows independent development and maintenance, enhancing flexibility and resilience [6]. Standardized interfaces further promote easy integration, regardless of the internal workings of each module [6].

3.5 Leveraging Composability in Systems Engineering

Composability in systems engineering is about building complex systems through well-defined, self-contained components that interact seamlessly. This design principle accelerates development, enhances maintainability, and scales efficiently. The following steps outline how to effectively leverage composability:

- Component Identification: By breaking down the system into distinct, self-contained elements, each enables a specific function, isolated from others to streamline development, testing, and maintenance. Clear identification clarifies the system's structure, function and simplifies interactions [23].
- Interface Definition: Establishing clear, standardized interfaces for each component ensures smooth integration and interaction. Standardized communication protocols allow components to work together without requiring insight into each other's internal mechanisms [24].
- Loose Coupling Design: Minimizing dependencies between components allows each part to operate independently. This design choice enhances modularity, simplifies updates, promotes parallel development, and boosts system resilience [25].

- Reuse and Integration Strategy: Creating a strategy to leverage existing components across projects optimizes efficiency. Reusing proven, well-tested modules accelerates development, reduces costs, and maintains consistency [26].
- Parallel Development Enablement: Loose coupling and modular design enable multiple components to be developed simultaneously. This parallel development reduces integration conflicts, shortens project timelines, and enhances team productivity [27].
- Testing and Validation: Thorough evaluation at both component and system levels ensures compatibility, reliability, and performance. Rigorous testing confirms that each module functions independently and integrates smoothly within the broader system [28].

These concepts can influence how to plan, design, and develop scalable and maintainable systems by encouraging a consideration of different approaches. Focusing on composability and modularity opens up new ways to think to help planning and development, making it easier to adapt and refine systems as needed.

Causality and composability are helpful approaches in effective systems engineering. Understanding how changes in one part of a system impact others (causality) and designing components that can be easily assembled and reconfigured (composability) enable more resilient, scalable, and maintainable system architectures.

These concepts drive more thoughtful planning, more explicit dependencies, and more efficient integration throughout the development lifecycle. Ultimately, a development strategy should explore diverse ways to uncover valuable insights, reducing the ambiguity often associated with complex projects. A more substantial and grounded perspective can be achieved by structuring analysis around causality and composability, making it more manageable to predict outcomes and adapt to changes.

Causality refers to understanding the cause-and-effect relationships within a system, while composability leverages modularity and standardization to optimize the integration of components. Applying these concepts together can lead to important gains like such as increased flexibility, smoother integration, and stronger scalability for example [29]:

- Improved System Design: By analyzing causal links, engineers can identify critical interactions and optimize designs to enhance overall system performance.
- Enhanced Predictability: A clear understanding of causality allows engineers to predict how changes in one part of the system will affect the whole, minimizing unexpected behaviors and optimizing performance.
- Failure Mode Identification: Causal analysis helps to detect underlying mechanisms that could lead to failures, enabling proactive mitigation strategies and improved reliability.
- Increased Flexibility and Adaptability: Modular, composable architectures enable rapid adaptation to new technologies and changing requirements with minimal disruption.

Leveraging the principles of causality and composability can be helpful for effectively designing and maintaining complex systems. Focusing on modularity and standardization simplifies system architecture, enhancing both resilience and scalability. Adopting standardized processes through composability significantly improves communication, streamlines knowledge sharing, and supports trade-off analysis among engineering development teams. These practices can enable clearer planning and design pathways, ensuring that projects remain adaptable and maintainable.

3.6 Causality, Composability, and Abstraction in Modern Systems Engineering

Causality and composability are valuable in contemporary systems engineering, forming the foundational framework for resilient, scalable, and adaptable system designs. By leveraging modularity and standardization, engineers can enhance performance, predict potential challenges, and streamline integration and maintenance. These form the basis for facilitating the development of intricate, high-performance systems that are easily reconfigurable and adaptable to evolving technological progress.

Composability inherently supports abstraction and hierarchy by promoting a clear understanding of how each component fits within the broader system. Abstraction allows engineers to focus on higher-level system behavior without getting lost in the details of individual components. At the same time, hierarchy enables structured layering, making complex systems more manageable and navigable [30].

Understanding system composability and causality provides a strong foundation for analyzing and navigating complex systems. This perspective not only improves the ability to learn and interpret system behavior but also guides engineers toward more focused problem-solving and informed decision-making. By applying these concepts, development efforts become more efficient, structured, and adaptable, ultimately supporting the creation of systems that are both resilient and responsive to evolving requirements.

References

1. Wagner, A. (1999). Causality in complex systems. *Biology & Philosophy, 14*(1), 83. https://doi.org/10.1023/a:1006580900476, Retrieved from: https://link.springer.com/article/10.1023/A:1006580900476
2. Jia, L., Cong, Z., Man, K. L., Guan, S.-U., Smith, J. C., & Yue, Y. (2022). *Cause of causal emergence: Redistribution of uncertainty and its critical condition.* arXiv (Cornell University). https://doi.org/10.48550/arxiv.2212.01551
3. Gao, G., Mishra, B., & Ramazzotti, D. (2017). *Causal data science for financial stress testing.* arXiv. https://doi.org/10.48550/ARXIV.1703.03076
4. Iwuozor, J. (2023, November 8). *What is composability and why should you care?* Retrieved from https://www.progress.com/blogs/what-composability-why-should-you-care

5. Pearl, J. (2009). *Causality: Models, reasoning, and inference* (2nd ed.). Cambridge University Press. https://bayes.cs.ucla.edu/BOOK-2K/neuberg-review.pdf
6. Langford, G. O. (2012). *Engineering systems integration: Theory, metrics, and methods* (pp. 75–90). https://library.oapen.org/bitstream/20.500.12657/40121/1/9781439852880.pdf. Retrieved from: https://www.researchgate.net/publication/346556661_Engineering_Systems_Integration_Theory_Metrics_and_Methods
7. Loftus, J. R., Russell, C., Kusner, M. J., & Silva, R. (2018). *Causal reasoning for algorithmic fairness.* arXiv (Cornell University). https://doi.org/10.48550/arxiv.1805.05859. Retrieved from https://arxiv.org/pdf/1805.05859
8. Gendron, G., Witbrock, M., & Dobbie, G. (2023). *A survey of methods, challenges and perspectives in causality.* arXiv (Cornell University). https://doi.org/10.48550/arxiv.2302.00293
9. Cox, L. A. (2021). Toward practical causal epidemiology. *Global Epidemiology, 3*, 100065. Retrieved from: https://pmc.ncbi.nlm.nih.gov/articles/PMC10446107/
10. Zanga, A., & Stella, F. (2023). *A Survey on Causal Discovery: Theory and practice.* arXiv (Cornell University). https://doi.org/10.48550/arxiv.2305.10032
11. Petersen, M., & van der Laan, M. J. (2014). Causal models and learning from data [review of causal models and learning from data. *Epidemiology, 25*(3), 418. Lippincott Williams & Wilkins. https://doi.org/10.1097/ede.0000000000000078
12. Schmidt, S., Lam, D., & Hayden, P. (2021). *Automatic knowledge extraction with human interface.* arXiv (Cornell University). https://doi.org/10.48550/arxiv.2104.04415
13. Szabó, R., & Vörös, A. (2021). *Towards formally analyzed cyber-physical systems.* arXiv (Cornell University). https://doi.org/10.48550/arxiv.2108.12773
14. DeTurris, D., & Palmer, A. (2018). Perspectives on managing emergent risk due to rising complexity in aerospace systems. *Insight, 21*(3), 80. Retrieved from: https://www.researchgate.net/publication/328959161_Perspectives_on_Managing_Emergent_Risk_Due_to_Rising_Complexity_in_Aerospace_Systems
15. Blomkamp, E. (2021). Systemic design practice for participatory policymaking. *Policy Design and Practice, 5*(1), 12. Retrieved from: https://www.tandfonline.com/doi/full/10.1080/25741292.2021.1887576#d1e213
16. Cotsaftis, M. (2009). What makes a system complex?—An approach to self organization and emergence. In *Understanding complex systems* (p. 49). Springer Nature. https://doi.org/10.1007/978-3-642-02199-2_3
17. Vainorenie, I. P., & Lemish, N. (2020). Multidisciplinary aspects of causality as an ontological and epistemological category. *Journal of History Culture and Art Research, 9*(1), 305. Retrieved from: https://www.researchgate.net/publication/340834188_Multidisciplinary_Aspects_of_Causality_as_an_Ontological_and_Epistemological_Category
18. Eckert, C., Clarkson, P. J., & Zanker, W. (2004). Change and customisation in complex engineering domains. *Research in Engineering Design, 15*, 1–21 (2004). https://link.springer.com/article/10.1007/s00163-003-0031-7
19. *The architecture of complexity: hierarchical systems.* (2024). Retrieved from: https://public.econ.duke.edu/~kdh9/SourceMaterials/ZurichPapers/SimonCh.8.pdf
20. Poss, R. (2013). *The essence of component-based design and coordination.* arXiv (Cornell University). https://doi.org/10.48550/arxiv.1306.3375
21. Li, J., & Kwauk, M. (2003). Exploring complex systems in chemical engineering—The multiscale methodology. *Chemical Engineering Science, 58*, 521. Retrieved from: https://doi.org/10.1016/s0009-2509(02)00577-8
22. Chauhan, K., Peltokorpi, A., Lavikka, R., & Seppänen, O. (2022). The monetary and non-monetary impacts of prefabrication on construction: The effects of product modularity. *Buildings, 12*(4), 459. https://doi.org/10.3390/buildings12040459

23. Bass, L., Clements, P., & Kazman, R. (2012). *Software architecture in practice.* Addison-Wesley. Retrieved from: https://www.researchgate.net/publication/22400112
24. Shaw, M., & Garlan, D. (1996). *Software architecture: Perspectives on an emerging discipline* (p. 97). Prentice Hall. Retrieved from: https://archive.org/details/softwarearchitec0000shaw_q2r0
25. Parnas, D. L. (1972). On the criteria to be used in decomposing systems into modules. *Communications of the ACM, 15*(12), 1053–1058. Retrieved from: https://www.researchgate.net/publication/200085877_On_the_Criteria_To_Be_Used_in_Decomposing_Systems_into_Modules
26. Szyperski, C. (2002). *Component software: Beyond object-oriented programming* (p. 224). Addison-Wesley. Retrieved from: https://archive.org/details/componentsoftwar0000szyp_p3o5
27. Kruchten, P. (2004). *The rational unified process: An introduction.* Addison-Wesley. Retrieved from: https://www.researchgate.net/publication/220018149_The_Rational_Unified_Process--An_Introduction
28. Myers, G. J., Sandler, C., & Badgett, T. (2011). *The art of software testing* (pp. 85–113). Wiley.
29. Lamb & Rhodes. (2007). Retrieved from: https://web.mit.edu/adamross/www/LAMB_INCOSE07.pdf
30. Maier, M. W., & Rechtin, E. (2009). *The art of systems architecting* (pp. 137–144). CRC Press.

4 Understanding Problem and Solution Spaces in Systems Engineering

In systems engineering, the interaction between problem and solution spaces is fundamental for effective project planning and development. The **problem space** represents the needs, goals, and requirements of stakeholders, providing a structured understanding of the issues that need resolution. In contrast, the **solution space** encompasses all viable design alternatives that meet the specified requirements, allowing for the configuration of different solutions as new information emerges.

The distinction is clear: the **problem space** defines the *what*, while the **solution space** defines the *how*. The problem space emphasizes understanding the problem, including needs, goals, dependencies, and constraints. Meanwhile, the solution space is where potential solutions are evaluated against the problem's specifications.

Systems engineering is described as a methodical approach to designing, realizing, and operating systems, with a strong emphasis on satisfying stakeholder needs within defined constraints. This description aligns with the dynamics of the problem and solution spaces: the problem space captures stakeholder requirements and limitations, and the solution space explores configurations that meet those demands [1].

4.1 Practical Problem Solving: Strategies and Techniques

A well-defined problem space is essential for effectively understanding the solution space. Recognizing whether the right problem is being addressed may seem logical, but it is often overlooked in problem-solving. This underscores the need to reframe problems to uncover unexpected solutions. Consider the following: [2]

F. A. Macias, *Project Development Simplified*, Synthesis Lectures on Engineering, Science, and Technology, https://doi.org/10.1007/978-3-032-01043-8_4

- **Challenge the Initial Framing**: Initial problem statements are often narrow or symptom-focused. Questioning assumptions and exploring different perspectives can uncover the true root cause.
- **Consider Different Levels of Abstraction**: Analyzing the problem from various levels of detail can reveal systemic issues when zoomed out and pinpoint specific challenges when zoomed in.
- **Involve Diverse Stakeholders**: Different stakeholders bring unique insights. Engaging with them helps identify blind spots and ensures a comprehensive understanding of the problem.
- **Analyze Positive Exceptions**: Identify situations where the problem did not occur. Understanding what differed in those scenarios can offer clues for potential solutions.
- **Test and Iterate**: Experimentation and iterative testing help refine problem understanding and solution effectiveness, echoing iterative approach in project management [3].

4.2 Exploring the Solution Space

The problem space encompasses the entire problem, including its context, stakeholders, requirements, and constraints. It defines both the current state and the desired future state. Conversely, the solution space includes all conceivable solutions to the problem as perceived at that point in time, from incremental improvements to innovation. Developers must conceptualize and evaluate various approaches within this space to address the needs outlined in the problem space [4].

A well-structured problem space allows for more efficient exploration of the solution space. In design, the solution space can be mapped using optimization algorithms, such as box and polytope optimization, which aim to maximize the volume of feasible designs. Design patterns further bridge these spaces by providing established templates that solve recurring issues across different scenarios.

4.3 Systems Engineering and Hierarchical Design

Understanding the relationship between these two spaces is crucial for effective systems planning. Progressing toward this goal fosters the development of practical, implementable solutions resilient to changing requirements. Moreover, hierarchical design methods can simplify complex systems by logically structuring requirements-defined components [4, 5].

When the problem space is clearly defined, it establishes a solid foundation for efficiently exploring the solution space. Clear boundaries, specific requirements, and well-understood constraints help narrow the search, reduce ambiguity, and focus on high-potential areas. This structured approach not only accelerates the discovery of optimal solutions but also minimizes wasted effort on paths that are unlikely to succeed [6].

Furthermore, exploring the solution space can drive effective concurrency strategies in software development. By understanding the structure of the components of the solution, developers can design systems that leverage multi-threading and distributed computing. This not only boosts performance but also optimally manages computational resources. For instance, identifying independent solution paths during exploration allows for simultaneous execution, improving throughput and reducing bottlenecks [6].

Therefore, the relationship is bidirectional: a well-defined problem space sharpens solution exploration, and iterative exploration uncovers opportunities for concurrency that enhance development efficiency and scalability [6].

To develop essential solutions, engineers must consider the feasible operating regions within the solution space, analyzing how adjustments to variables affect the solution's effectiveness. This cognitive framework, resulting from the combination of analysis, synthesis, and evaluation, remains foundational in systems design, supporting iterative learning and striving for continuous improvement.

Ultimately, understanding the concepts of problem space, solution space, open systems, and closed systems is essential for analysis to achieve successful engineering outcomes. Applying systems theory provides a structured way to articulate these elements, enhancing clarity and effectiveness in project execution [4].

4.4 The Role of Abstraction in Problem and Solution Spaces

Abstraction is a crucial concept in both computer science and design thinking, serving as a powerful tool for managing complexity by reducing intricate problems to their essential elements. In systems engineering, abstraction enables practitioners to generalize solutions from one problem to others with similar structures, enhancing efficiency and adaptability [7].

High levels of systems thinking allow engineers to grasp the broader scope of a project, both its functional and conceptual aspects, without becoming entangled in minute details. This macro-level understanding is essential for effective design and problem-solving in complex systems [8].

Understanding both problem and solution spaces is critical when developing sophisticated systems. Problem spaces represent the conditions, constraints, and challenges that define a system's requirements, while solution spaces encompass the range of potential designs and optimizations that address these requirements [9]. Computational thinking skills such as abstraction, pattern recognition, and modeling help navigate these spaces [7].

These skills allow engineers to identify hierarchical dependencies, causality, and composability, which are key to understanding how individual components interact within a larger system [10]. Through abstraction, systems engineers can distill complex problems into manageable concepts, identify recurring patterns, and develop generalized solutions that are applicable across diverse scenarios [11].

This simplification enables a more holistic understanding of systems, including their components and interdependencies, resulting in more resilient and flexible designs [10]. In addition, abstraction facilitates optimization in the solution space, often represented through box and polytope optimization algorithms, which strive to maximize the volume of feasible design solutions [12].

One of the most challenging aspects of development is accurately defining the requirements. Good design is not just solving a problem effectively, but solving the right one. This concept emphasizes the importance of accurately understanding user needs from the outset. However, requirements defined abstractly, detached from real-world contexts, are often flawed and misaligned with actual user behaviors and expectations. Similarly, requirements derived solely from asking users what they need are frequently unreliable, as people often struggle to articulate their needs or foresee how they will interact with a system [13].

Contextual inquiry by observing stakeholders in their natural environments as they interact with their operational setting is considered one of the most effective methods for gathering project requirements [14]. This approach uncovers real-world challenges, inefficiencies, and hidden needs that are often challenging to identify through traditional methods [15]. Grounding project requirements in these direct observations helps ensure that solutions are more intuitive, usable, and closely aligned with actual needs.

Abstraction plays a critical role in making sense of the raw data collected during contextual inquiry. When designers observe complex user interactions and varying workflows, abstraction allows them to distill these observations into generalized patterns and essential concepts [7]. This process of simplifying and structuring information helps designers to recognize commonalities across different scenarios and user behaviors, even when surface-level actions differ.

For instance, watching how different stakeholders perform tasks in a hospital can show unique processes related to their roles such as nurses, doctors, and administrative staff might handle the same system in different ways. Abstraction enables designers to eliminate these specific role details and focus on essential system interactions, like data retrieval, input, and communication flows. These broad abstractions subsequently guide the development of adaptable, scalable solutions that support various user types without needing significant customization [13].

Additionally, abstraction facilitates the correlation of user interactions with system design by modeling workflows and information flows. By operating at a higher level of generalization, developers can visualize the interactions of various system components without becoming overwhelmed by unnecessary details. This approach helps clarify the connections between user actions and system responses, simplifying the identification of optimization opportunities and improving the user experience [16].

4.5 Understanding Problem and Solution Spaces

In systems engineering, the concepts of problem space and solution space are fundamental to addressing design challenges effectively. The problem space represents the domain where stakeholders' needs, goals, and requirements are identified and analyzed. This forms the foundation for understanding the issues that require resolution [4]. Conversely, the solution space encompasses all feasible design alternatives that meet these defined requirements. This space is dynamic, evolving as new information becomes available, allowing engineers to explore various configurations [17].

These two spaces are interconnected through continuous analysis and iteration, moving back and forth as solutions are tested and refined based on insights gained. Effective systems engineering requires a deep understanding of both spaces to bridge the gap between needs and solutions.

4.5.1 The Role of Abstraction and Hierarchy

Abstraction is critical in managing complexity. It enables engineers to focus on essential system aspects while ignoring less relevant details, simplifying complex problems into more manageable components. This is especially valuable when requirements are ambiguous or incomplete, as abstraction allows progress with partial information [18].

Hierarchy further structures complexity by organizing system elements into different levels of abstraction. High-level concepts are progressively refined into more detailed designs. This hierarchical method simplifies analysis and supports modular development, allowing components to be independently designed and tested without disrupting the broader system [19].

Together, abstraction and hierarchy empower engineers to navigate complex problem spaces and efficiently explore solution spaces, enhancing both analysis and design processes.

4.6 Composability and Design Patterns

Composability refers to the process of combining simpler parts into more complex systems while maintaining their functionality and predictability. This is crucial for modular designs where individual components interact seamlessly within a system. Effective composability demands a clear understanding of the causal relationships between components to avoid unintended effects during integration [20]. Design patterns act as bridges between problem and solution spaces. They are established templates that offer proven solutions to recurring design problems. By leveraging these patterns, engineers accelerate development, enhance reliability, and ensure consistency across systems [21].

4.6.1 Business Analytics as a Framework

Business analytics in a systems engineering context involves the structured use of data, quantitative methods, and analytical tools to support informed decision-making across complex operations. This discipline plays a critical role in identifying performance gaps, evaluating process efficiency, and guiding strategic planning by integrating data-driven insights with engineering judgment and organizational objectives. When applied effectively, business analytics helps system engineers assess alternatives, prioritize resources, and validate design or operational changes within the broader system lifecycle. This approach supports continuous improvement by providing a clear, evidence-based understanding of how systems perform in real-world environments. Business analytics provides a structured approach for formalizing problem and solution spaces through:

- Descriptive Analytics: Understanding the current state and identifying trends [22].
- Diagnostic Analytics: (2023). Analyzing the causes behind observed trends [23].
- Predictive Analytics: Forecasting future outcomes based on historical data [24].
- Prescriptive Analytics: Optimizing solutions through simulation and decision support [22].

These techniques enhance abstraction by summarizing complex data into high-level insights and support hierarchical exploration through drill-down analysis. This structured methodology facilitates more transparent decision-making and more efficient solution design.

4.6.2 Systems Thinking: Holistic Problem-Solving

Systems thinking provides a holistic framework for understanding interdependencies within problem and solution spaces. It emphasizes recognizing feedback loops and mitigating unintended consequences. Systems thinking introduces the Cynefin framework for categorizing problems as simple, complicated, complex, or chaotic, guiding the choice of analysis methods accordingly [25]. In the solution space, systems thinking ensures that designs are integrated smoothly into broader systems, preventing conflicts and supporting sustainable development.

Abstraction and dependency analysis are critical for dissecting complex systems. Dependency analysis reveals hidden relationships, enabling engineers to forecast the effects of changes. This foresight is vital for avoiding integration conflicts and upholding system integrity.

By concentrating on core components and their interactions, abstraction simplifies complexity in both the problem and solution spaces. It also facilitates the exploration of design alternatives without excessive detail. Effective systems engineering relies on understanding the relationship between the problem space and the solution space. Through abstraction, hierarchy, composability, business analytics, and systems thinking, engineers can manage complexity, anticipate challenges, and create resilient, adaptable systems. Comprehending these concepts empowers engineers to deliver designs that fulfill stakeholder needs and adapt to changing conditions [3].

References

1. Shishko, R., Aster, R. W., Chamberlain, R. G., McDuffee, P., Pieniazek, L., Rowell, T., Bain, B., Cox, R. I., Mooz, H., & Polaski, L. (1995). *NASA systems engineering handbook* (pp. 172–188).
2. Wedell-Wedellsborg, T. (2017). *Decision making and problem solving are you solving the right problems?* https://hbr.org/2017/01/are-you-solving-the-right-problems
3. Wysocki, R. K. (2009). *Effective project management: Traditional, agile, extreme* (pp. 195).
4. Salado, A. (2021). A systems-theoretic articulation of stakeholder needs and system requirements. *Systems Engineering, 24*(2), 83. Wiley. https://www.researchgate.net/publication/348458206_A_systems-theoretic_articulation_of_stakeholder_needs_and_system_requirement
5. Vencel, L., Sweetman, E., & Lante, N. (2018). Composing complex capability: The stakeholder driven development framework. *INCOSE International Symposium, 28*(1), 1519. Wiley. https://doi.org/10.1002/j.2334-5837.2018.00565.x
6. Joglekar, N. R., Yassine, A., Eppinger, S. D., & Whitney, D. E. (2001). Performance of coupled product development activities with a deadline. *Management Science, 47*(12), 1605. https://www.researchgate.net/publication/264372329_Performance_of_Coupled_Product_Development_Activities_with_a_Deadline
7. Wing, J. M. (2006). Computational thinking. *Communications of the ACM, 49*(3), 33–35. https://www.researchgate.net/publication/274309848_Computational_Thinking
8. Checkland, P. (1999). *Systems thinking, systems practice* (pp. 5–13). Wiley. https://archive.org/details/systemsthinkings00chec/page/n1/mode/2up?q=systems+thinking
9. Simon, H. A. (1996). *The sciences of the artificial* (pp. 27). MIT Press. https://monoskop.org/images/9/9c/Simon_Herbert_A_The_Sciences_of_the_Artificial_3rd_ed.pdf
10. Jackson, M. C. (2019). *Critical systems thinking and the management of complexity.* Wiley. https://learning.oreilly.com/library/view/critical-systems-thinking/9781119118374/c05.xhtml#head-2-27
11. Larman, C. (2004). *Applying UML and patterns: An introduction to object-oriented analysis and design and iterative development* (pp. 140–161). Prentice Hall.
12. Boyd, S., & Vandenberghe, L. (2004). *Convex optimization* (p. 16). Cambridge University Press.
13. Norman, D. (2013). *The design of everyday things* (pp. 229). Basic Books.
14. Beyer, H., & Holtzblatt, K. (1998). *Contextual design: Defining customer-centered systems* (pp. 146–147). Morgan Kaufmann.

15. Goodman, E., Kuniavsky, M., & Moed, A. (2012). *Observing the user experience: A practitioner's guide to user research* (pp. 84). Morgan Kaufmann
16. Preece, Sharp, & Rogers. (2015). Pp. 179. https://www.google.com/books/edition/Interaction_Design/b-v_6BeCwwQC?hl=en&gbpv=1
17. Goel, V., & Pirolli, P. (1992). The structure of design problem spaces. *Cognitive Science, 16*(3), 395. https://www.sciencedirect.com/science/article/pii/036402139290038V
18. Öksüz, E. B., & Çağdaş, G. (2020). An assessment method for a designerly way of computational thinking. *A/Z ITU Journal of Faculty of Architecture, 17*(2), 199. https://doi.org/10.5505/itujfa.2020.86729. https://www.az.itu.edu.tr/index.php/jfa/article/view/337/227
19. Larsson, D. T., Maity, D., & Tsiotras, P. (2021). *Information-theoretic abstractions for resource-constrained agents via mixed-integer linear programming* (Vol. 66, p. 1). https://arxiv.org/abs/2102.10015
20. Wagner, N. (2024). Comparing the complexity and efficiency of composable modeling techniques for multi-scale and multi-domain complex system modeling and simulation applications: A probabilistic analysis. *Systems, 12*(3), 96. Multidisciplinary Digital Publishing Institute. https://www.researchgate.net/publication/378988833_Comparing_the_Complexity_and_Efficiency_of_Composable_Modeling_Techniques_for_Multi-Scale_and_Multi-Domain_Complex_System_Modeling_and_Simulation_Applications_A_Probabilistic_Analysis
21. Lesko, T. M. (2019). Teaching a thorough, repeatable design process for ambiances in architecture. In *SHS web of conferences* (Vol. 64, p. 2006). EDP Sciences. https://www.researchgate.net/publication/335399136_Teaching_a_thorough_repeatable_design_process_for_ambiances_in_architecture
22. Wissuchek, C., & Zschech, P. (2024). Prescriptive analytics systems revised: a systematic literature review from an information systems perspective. In *Information systems and e-business management*. Springer Science+Business Media. https://www.researchgate.net/publication/383566175_Prescriptive_analytics_systems_revised_a_systematic_literature_review_from_an_information_systems_perspective
23. Soldic Aleksic, J., Chroneos Krasavac, B., & Karamata, E. (2019). Business analytics: new concepts and trends. *Management: Journal of Sustainable Business and Management Solutions in Emerging Economies, 25*. https://doi.org/10.7595/management.fon.2019.0013. https://www.researchgate.net/publication/334679496_Business_analytics_new_concepts_and_trends
24. Balkan, S., & Demirkan, H. (2015). Teaching predictive model management in MIS classrooms: A tutorial. In *Communications of the association for information systems* (Vol. 37). Association for Information Systems. https://aisel.aisnet.org/cgi/viewcontent.cgi?article=3893&context=cais
25. Dettmer, H. W. (2011). Systems thinking and the Cynefin framework. https://static1.squarespace.com/static/578d0f8459cc6877481865ef/t/57ec5284579fb363a5b9a18c/1475105417186/Systems-Thinking-and-the-Cynefin-Framework-Final.4.pdf

5 Bricolage

Bricolage, a concept introduced by anthropologist Claude Lévi-Strauss, captures the art of solving problems with whatever is at hand. It is about piecing together ideas, tools, and materials from different places to create something new. This kind of inventive problem-solving is useful in engineering, where design, adaptability, and practical ingenuity are useful in problem-solving. In systems engineering, Bricolage aligns well with concepts of abstraction and hierarchy. When engineers face challenges in a project development, they often abstract problems to higher levels to minimize complexity and identify core system functions and interactions [1].

In systems engineering, Bricolage aligns well with concepts of abstraction and hierarchy. When engineers face challenges in a project development, they often abstract problems to higher levels, stripping away complexity to identify core functions and interactions. The application of abstraction helps to see the broader structure of the system, enabling creative analysis of existing components and defining what needs to be added, much like Bricolage, to solve problems [2].

The hierarchical nature of system design enables engineers to address issues at different layers, applying appropriate solutions at lower levels without disrupting the entire architecture. When conventional methods fall short, a process such as bricolage encourages inventive thinking, leveraging the modularity and scalability built into hierarchical design [3].

In The Genius of the Tinkerer, Steven Johnson connects bricolage to how ideas evolve, especially in engineering, where progress often comes from reusing and reimagining what already exists. However, for this kind of improvisation to be effective, it must work within some form of structure, clearly defined or loosely understood. This balance between creativity and order is essential in systems engineering, where outcomes must

F. A. Macias, *Project Development Simplified*, Synthesis Lectures on Engineering, Science, and Technology, https://doi.org/10.1007/978-3-032-01043-8_5

remain functional and aligned with broader goals. Without that structure, innovation risks becoming aimless or ineffective [4].

5.1 Bricolage in Systems Engineering

Bricolage as a process can be relevant in systems engineering, where adaptability, iterative development, and resource optimization are critical for managing constraints like limited budgets, timelines, and system requirements. Examples of bricolage applications in systems engineering include system integration, software reuse, and information reuse and integration as follows:

System Integration: The benefit of Bricolage in systems engineering is in defining the utilization of existing resources to construct a cohesive system. This often leverages service-oriented architecture (SOA) principles to help streamline integration tasks, enhance modularity, and improve cost efficiency [5].

Service-Oriented Architectures (SOA) are still relevant today, especially in enterprise environments where system integration, modularity, and interoperability are important requirements. While micro-services are favored for their simplicity and independence, SOA continues to be used in managing complex, distributed systems, enabling service reuse and smooth integration while reducing development time and operational costs. SOA's design principles align well with cloud computing and IoT, reinforcing its lasting relevance [6].

Resource bricolage involves creatively repurposing available resources to address challenges and overcome constraints. Integrating the modular structure of service-oriented architecture (SOA) with this kind of resourcefulness enables systems engineers to develop flexible, cost-effective solutions that adjust to evolving requirements without needing significant new development. Using existing components and services makes integration easier and lowers costs [7].

Software Reuse: A structured software reuse strategy is a form of Bricolage, as evident by reusing existing designs and code components to address new problems. This strategy could save time and money as this approach makes the most of available resources, turning limitations into opportunities. For example, the NASA Constellation Program focused on reusing software components to cut costs and meet tight schedules, showcasing how software reuse can be effective in large-scale engineering projects [8]. Systematic software reuse has been recognized as a strategy for introducing products quickly to market while maintaining quality and lowering costs, especially in competitive development environments challenged by budgetary restrictions [9].

Information Reuse and Integration: Maximizing existing data and system architectures minimizes redundancy and enhances decision-making within large-scale system developments. In systems engineering, iterative development enables continuous refinement of system designs as new information and insights become available by applying structured reuse of components and iterative enhancements. This can reduce risks while optimizing system reliability, cost-efficiency, and adherence to project schedules. Reusing existing components leverages already tested and validated elements, lowering the chances of unexpected failures [10].

Bricolage in Systems Engineering

Despite its benefits, Bricolage has some challenges, such as maintaining compatibility between existing processes and new developments. Integrating reusable components must satisfy system requirements such as strict quality and efficiency requirements. In addition, as is typical with any development, adding new functions can introduce complexity and subsequent cost to development. Adding new infrastructure capabilities often demands specialty engineering, a key component of systems engineering that complements technical activities as required to deliver a project by leveraging custom technical expertise.

In the Harvard Business Review article "Artificial Intelligence for the Real World" (January–February 2018), Thomas H. Davenport and Rajeev Ronanki present framework for implementing AI in business, cautioning against overly ambitious "moonshot" projects and advocating for incremental, value-driven applications. Integrating cognitive project which will require a fair amount of specialty engineering as the most challenging obstacle [11].

Specialty engineering focuses on structured methods for reliability and precision, while Bricolage focuses on solving problems creatively with whatever is available. Together, they help systems engineers mitigate challenges and adapt in fast-changing environments [12].

A notable example of integrating specialty engineering with Bricolage in systems engineering is the Apollo 13 mission. Following an oxygen tank explosion, the crew faced rising carbon dioxide levels due to incompatible lithium hydroxide canisters between the Command Module (CM) and the Lunar Module (LM). NASA engineers in Mission Control in Houston, Texas, devised an improvised adapter using available materials like plastic covers and duct tape found on the spacecraft to fit the square CM canisters into the round LM receptacles.

Developed under significant pressure, this solution exemplifies how structured engineering practices combined with creative problem-solving can effectively address unforeseen challenges [13].

Despite its advantages, bricolage presents challenges in ensuring compatibility between previously used processes and new developments. The integration of reusable components must meet strict quality and efficiency requirements, increasing the complexity and cost of development. Furthermore, integrating new capabilities into existing infrastructures often requires specialty engineering, which relies on precise technical expertise [14].

While specialty engineering involves structured methodologies, bricolage fosters creative problem-solving. The intersection of these two approaches allows systems engineers to effectively navigate unexpected challenges. This fusion is instrumental in dynamic fields where adaptability is crucial for success [15].

5.2 Hierarchical Abstraction and Complexity in Systems Engineering

Hierarchical abstraction manages complexity in systems engineering by organizing information into structured layers that define the system's functions and interdependencies. This layered approach establishes a modular architecture that describes the system, helping to improve scalability, maintainability, and fault isolation while supporting efficient system optimization and the seamless integration of new components [16]. Consider the following:

- **Reducing Complexity**: Hierarchical structuring divides a system into manageable layers, making it easier to conceptualize, develop, and maintain. This simplification is linked to the principle of requisite variety in Computer Science, which states that to effectively regulate a system, the regulator must have a sufficient range of actions to counteract the various disturbances it may face. This ensures the system's internal state remains as close as possible to the desired goal state. By enhancing clarity and control over the system's components, this approach simplifies the management of complexity [17].
- **Application of Abstraction**: Emphasizing core system functionality while filtering out unnecessary details enhances clarity and supports more effective decision-making [18].
- **Logical Layers with Defined Relationships**: A structured approach establishes explicit dependencies, ensuring that changes in one layer have predictable and controlled effects on others. A functional analysis should enable the comprehensive and logical completion of the design, development, and system definition process. It involves identifying top-level requirements and then partitioning them to a second level and on down as needed to define overall functionality [19].

The relational dependencies between components emerge by organizing the system into structured layers, each representing a specific function contributing to the system's overall purpose. The intent is to mitigate complexity into more straightforward terms by isolating subsystems, supporting focused development, and providing a more precise mapping of dependencies. By structuring these components hierarchically, engineers can address each layer independently to fully address functionality while maintaining coherence of the overall system design. Additionally, hierarchical abstraction promotes the reuse of components since functional relationships are better understood, enabling scalable expansion and easier integration of new technologies [20].

5.2.1 Case Study: Apollo 13 and the Role of Bricolage

As highlighted earlier, the Apollo 13 mission offers a prime example of Bricolage. After an oxygen tank explosion severely damaged the Apollo spacecraft, NASA engineers were forced to develop a survival plan with only the materials available onboard the spacecraft. Engineers understood that a buildup of carbon dioxide was quickly putting the Apollo crew in danger. The challenge was to recreate an alternative version of the failed system. The concern is how an alternative approach will solve the problem, requiring quick thinking and resourcefulness. To achieve a solution, engineers repurposed duct tape and tubing to construct a makeshift CO_2 filter, effectively reducing the buildup of carbon dioxide and ensuring the crew's survival.

This approach shows the benefits of adaptability and resourcefulness when dealing with challenging situations, and the case illustrates [21]:

- **Bricolage**: Engineers creatively utilized available materials to address a critical problem. This method, known as resource bricolage, involves reconstructing, integrating, and repurposing existing resources to find effective solutions [22].
- **Abstraction**: Engineers prioritized critical survival tasks by breaking down complex system failures into manageable parts. Each task makes full use of the existing abstractions; moreover, it only generates necessary abstractions that are conducive to the new task [23].
- **Hierarchies**: The organized design of the spacecraft's systems allowed engineers to concentrate on life support while maintaining navigation and communication capabilities. Engineers understood the relationships of all complex systems within the spacecraft and identified stable processes based on the overall hierarchical organization, which facilitated the development of alternative solutions [24].

The Apollo 13 mission illustrates how bricolage, abstraction, and hierarchical structuring work together to facilitate effective problem-solving under extreme constraints.

5.3 Applications of Bricolage in Systems Engineering

Bricolage thrives in environments where engineers think creatively to overcome challenges. In contrast to traditional methods, which depend on predefined tools and rigid processes, Bricolage emphasizes practical experimentation and improvisation. Additionally, this approach leverages available resources to solve problems in innovative ways, which has proven to be highly effective in systems development. Similarly, it showcases adaptability and resourcefulness, particularly when responding to unexpected challenges or working with limited resources.

The usefulness of Bricolage becomes apparent where rapid prototyping, iterative design, and real-time problem-solving are needed. From space missions like Apollo 13 to disaster recovery scenarios and agile software development, Bricolage enables engineers to craft solutions swiftly and effectively using whatever is on hand. This approach reduces dependency on ideal conditions and fosters innovation in complex, high-pressure environments.

The concept of Bricolage thrives on adaptability, resourcefulness, and improvisation, often compared to qualities that are characteristic of human-centric. As advancements in AI continue to evolve, there is growing interest in understanding whether these technologies can replicate Bricolage, such as features like adaptability in problem-solving [25].

Modern AI systems, particularly those powered by machine learning and reinforcement learning, are beginning to demonstrate elements of improvisation by learning from dynamic environments and adjusting strategies based on real-time data. However, Bricolage requires not just adaptation but the ability to repurpose tools and approaches in unique ways that are different from predefined parameters [25].

However, it remains a challenge for AI, as it is often limited by its programmed logic and the data it is trained on to perform specific tasks. Enhancing its capacity for creative problem-solving may require advancements in contextual understanding, real-time decision-making, and multi-domain integration. As these technologies progress, the gap between conventional automation and flexible, systems-engineering approaches may narrow, resulting in more resilient and adaptable system designs [26].

5.4 Bricolage as a Catalyst for Abstraction

Bricolage plays a crucial role in abstraction by transforming complex problems into manageable components through creative problem-solving. The process of Bricolage fosters abstraction in the following ways:

- Emergent Properties: Bricolage often results in emergent system properties, where interactions among repurposed components create new functionalities beyond their original design [27].
- Incremental System Evolution: By leveraging existing resources, bricolage enables systems to evolve iteratively, allowing engineers to apply abstraction in modifying only specific layers without redesigning entire structures [28].
- Pattern Recognition: The improvisational nature of bricolage helps engineers recognize patterns in problem-solving, aiding in the abstraction process by identifying reusable design elements [29].
- Knowledge Integration: Bricolage supports the integration of knowledge from various disciplines, reinforcing abstract thinking and interdisciplinary problem-solving [4].

5.5 Hierarchical Analysis of Bricolage

Bricolage operates at different hierarchical levels, from component-level modifications to system-wide reconfigurations. Hierarchical abstraction structures this process, enabling effective system development through:

- Multi-Layered Adaptation: Bricolage allows for adaptability at various hierarchical levels. Tactical bricolage involves modifying subcomponents, while strategic bricolage reshapes workflows [30].
- Layered Decision-Making: Different system levels require different bricolage applications. Engineering teams may apply bricolage at the physical hardware level, while management may use bricolage for process optimization [31].
- Information Flow and Modularity: Bricolage supports hierarchical abstraction by enabling modular design principles where system elements interact through well-defined interfaces, ensuring system coherence [32].

5.6 Case Study: Bricolage in Aerospace System Failures

Bricolage has been instrumental in aerospace system adaptations and emergency problem-solving, as seen in the following examples:

- Apollo 13 CO_2 Filter Adaptation: NASA engineers created an improvised solution using available spacecraft materials, highlighting bricolage's role in abstraction and hierarchical structuring [21].
- International Space Station (ISS) Maintenance: Astronauts rely on bricolage solutions for system repairs, repurposing tools and components to adapt to unexpected failures [33].

- Space Shuttle Challenger and Columbia Missions: While bricolage solutions have been crucial for system resilience, these disasters highlight the limitations of improvisation when structural failures occur without hierarchical safeguards [34].

Bricolage principles are being adopted across various fields to enhance innovation and adaptability. Abstraction plays a crucial role in simplifying complex problems into manageable steps. For abstraction to be practical, the simplified actions must be more straightforward to execute than the original problem, ideally requiring no additional search or planning. A well-designed abstraction removes unnecessary details while preserving the core functionality, ensuring that each step can be efficiently completed. In artificial intelligence, machine learning models utilize bricolage by repurposing existing datasets and algorithms, applying abstraction to streamline problem-solving and enhance efficiency [35].

In cybersecurity, the unpredictable nature of cyber threats has led security teams to embrace bricolage-driven strategies, repurposing existing tools in real time to counter emerging risks. While bricolage contrasts with originality, it is increasingly recognized as a core element of digital culture and a catalyst for its rapid evolution [36].

Similarly, urban planning incorporates bricolage by blending traditional and modern infrastructure elements, promoting scalable and sustainable city development. These examples illustrate how bricolage enables resourceful problem-solving and flexible adaptation across technical, digital, and physical environments [37].

5.7 Bricolage as a Catalyst for Abstraction

Bricolage is a practical approach in systems engineering that helps break down complex problems into smaller, manageable parts. This method leverages existing system components and resources to effectively address challenges, thereby making design and implementation more efficient. One significant benefit is the emergence of new system properties, where combining repurposed components can lead to unexpected functionalities that go beyond the original intent of the developers [38]. This also supports incremental system evolution by leveraging existing resources, enabling systems to evolve iteratively and allowing engineers to apply abstraction in modifying specific layers without redesigning entire structures [28].

The flexible nature of bricolage facilitates pattern recognition, enabling engineers to identify reusable design elements that streamline problem-solving [39]. Additionally, bricolage facilitates the integration of knowledge from diverse disciplines, thereby enhancing abstract thinking and promoting interdisciplinary problem-solving [4].

At different hierarchical levels, bricolage functions through a structured process of adaptation and decision-making. Enabling a multi-layered adaptation across these levels, with tactical bricolage focusing on subcomponents and strategic bricolage reshaping entire workflows [30].

Layered decision-making reflects the application of bricolage at both physical hardware levels and process optimization. Moreover, bricolage supports hierarchical abstraction by promoting modular design principles, where system elements interact through well-defined interfaces, ensuring coherence and stability. These elements collectively illustrate how bricolage simplifies complex systems and enhances abstraction by effectively structuring and integrating diverse components [32].

5.8 Bricolage as a Catalyst for Abstraction

In systems engineering, bricolage serves as a powerful method for driving abstraction by breaking down complex problems into manageable, practical components. Through creative problem-solving and the repurposing of existing system elements, bricolage enables engineers to abstract critical system features without requiring complete redesigns. One key outcome of this process is the emergence of new system properties, unintended functionalities that arise from the interactions of repurposed components [40].

This phenomenon demonstrates how bricolage can drive innovative solutions that exceed initial design specifications. Bricolage also supports incremental system evolution by effectively utilizing existing resources. This approach enables targeted modifications within specific system layers, allowing for efficient abstraction without disrupting the overall architecture [28].

Recognizing patterns is a key aspect of abstraction in engineering design. The act of spotting and reapplying standard design features not only simplifies problem-solving but also improves the consistency of systems. A clear illustration of this method is the application of design patterns, which act as reusable micro-architectural elements that aid in the development of the entire system's architecture [41].

Additionally, the integration of cross-disciplinary knowledge is crucial in systems engineering. A holistic approach that spans multiple domains enables the development of systems that are resilient, adaptable, and aligned with the diverse needs of various stakeholders. Such cross-disciplinary collaboration is essential for tackling the complex challenges inherent in modern technological solutions [42].

5.9 Bricolage Applications

Bricolage operates effectively across multiple hierarchical levels within systems engineering, from fine-tuning individual components to reconfiguring large-scale systems. This layered structure facilitates methodical abstraction, optimizing both design and implementation. At its core, bricolage supports multi-layered adaptation, allowing modifications at different system levels to occur independently. Tactical bricolage hones in on individual subcomponents, ensuring optimal performance at the micro-level. Conversely, strategic bricolage focuses on broader workflow adjustments, reshaping entire processes to enhance efficiency and adaptability [30].

Layered decision-making amplifies the effectiveness of bricolage within hierarchical abstraction. Different system layers require unique bricolage strategies: engineering teams may employ bricolage at the hardware or software component level to optimize technical performance. At the same time, management applies bricolage principles to enhance organizational processes. This structured, multi-layered approach ensures coherent system development, allowing modifications at one level to occur seamlessly without disrupting functionality at others [3].

Bricolage also supports hierarchical abstraction through its emphasis on information flow and modularity. By encouraging modular design principles, bricolage ensures that system components interact seamlessly through well-defined interfaces. This modularity not only simplifies the integration of new elements but also maintains system coherence, allowing for efficient scaling and reconfiguration as requirements evolve.

Through these mechanisms, bricolage serves as a critical enabler of abstraction, driving both innovation and stability in complex systems engineering projects [32]. Bricolage enhances systems engineering by fostering adaptability, resourcefulness, and creative problem-solving. When combined with hierarchical abstraction, it enables the development of structured yet flexible systems.

This integration enables scalable, efficient, and resilient solutions across various fields, including aerospace and artificial intelligence. Understanding the interplay between bricolage and hierarchical structuring is essential for addressing modern engineering challenges and optimizing system design.

Bricolage and abstraction are both tools for managing complexity, just at different levels. Bricolage works from the bottom up: it starts with available parts, ideas, or solutions and combines them creatively to solve problems. Abstraction works top-down, simplifying complexity by focusing on essential features and ignoring irrelevant details.

What ties them together is pattern recognition. Through bricolage, engineers and designers discover what works in practice. Over time, these improvised solutions reveal recurring patterns, which can then be abstracted. For example, a quick fix that's reused across multiple projects may eventually become a formal design pattern or a reusable module.

Although bricolage and abstraction may seem conceptually distinct, with bricolage rooted in improvisation and abstraction grounded in formal simplification, they are deeply interconnected in practice. Bricolage involves creating solutions by repurposing available materials and ideas, often within the limitations of a given environment. This creative problem-solving does not eliminate the need for structure. Instead, it works within or around existing structures to find new configurations that address immediate needs [2].

Abstraction, in contrast, is a cognitive and design strategy for managing complexity. It works by focusing on essential features while setting aside less relevant details [28]. In systems engineering, abstraction supports modular thinking, layered design, and the development of scalable architectures [32]. When considered together, bricolage can be seen as a generative process that informs and drives abstraction. Solutions that emerge through bricolage often reveal repeatable patterns, which can later be distilled into formalized design concepts or reusable components [39].

This connection becomes especially clear in the history of innovation, where many engineering breakthroughs come not from entirely new inventions but from the creative recombination of existing tools and ideas. Once proven effective, these recombinations are often formalized into frameworks, protocols, or patterns. In this way, bricolage precedes abstraction. They produce the practical insights and working examples that abstraction then captures and generalizes [4].

Bricolage also facilitates incremental system evolution. Engineers frequently respond to challenges by adjusting specific components, rather than overhauling entire systems. This approach aligns with the perspective on complex systems as hierarchical and decomposable structures [28]. The concept of emergent properties further supports this idea, where the interactions among components lead to new system behaviors that cannot be predicted from the individual parts alone [27].

In interdisciplinary contexts, bricolage supports knowledge integration by enabling engineers to draw insights from diverse fields. This process fosters the development of abstractions that are broadly applicable and grounded in real-world problem-solving. This dynamic is particularly valuable in systems engineering, where technical, organizational, and human considerations must be addressed simultaneously [4, 30].

Bricolage does not merely support abstraction. It helps create it. Through hands-on adaptation, the recognition of functional patterns, and the continuous refinement of ideas, bricolage acts as a pathway to abstraction. Together, these two processes form a feedback loop that enables innovation in the design and evolution of complex systems.

References

1. Markham, A. (2017). *Bricolage: A keyword in remix studies.* Retrieved from https://annettemarkham.com/2017/01/bricolage-a-keyword-in-remix-studies/#:~:text=Origin. From Levi Strauss clarifies that.
2. Lévi-Strauss, C. (1966). *The savage mind* (G. Weidenfeld & Nicolson, Trans.). University of Chicago Press. (Original work published 1962) (pp. 17, 21, 29). Retrieved https://www.google.com/books/edition/The_Savage_Mind/JI6GVFbP9hAC?hl=en&gbpv=1&bsq=bricolage
3. Büscher, M., Gill, S. P., Mogensen, P., & Shapiro, D. (2010). *Landscapes of practice: Bricolage as a method for situated design.* Retrieved from https://www.researchgate.net/publication/220169078_Landscapes_of_Practice_Bricolage_as_a_Method_for_Situated_Design

4. Johnson, S. (2010). *The genius of the tinker. The Wall Street Journal.* Retrieved from https://www.wsj.com/articles/SB10001424052748703499604575512751348816216?mso ckid=07ba4fa575b66e513f905c3974c36f76.
5. What is SOA (Service-Oriented Architecture)? (2024). Retrieved from: https://aws.amazon.com/what-is/service-oriented-architecture/?utm_source=chatgpt.com
6. IBM, What is service-oriented architecture (SOA)? (2024). Retrieved from: https://www.ibm.com/think/topics/soa?utm_source=chatgpt.com
7. X., Zhang, Z., & Chun, D. (2025). Resource bricolage, digital transformation, and business model innovation: Based on the conditional process analysis of entrepreneurship. *Systems, 13*(5), 355. Retrieved from https://www.mdpi.com/2079-8954/13/5/355?utm_source=chatgpt.com
8. Atkinson, D. J. (2005). Software strategy for reuse study: Final report. *National Aeronautics and Space Administration, Exploration Systems Mission Directorate.* https://doi.org/10.13140/2.1.3399.3604. Retrieved from: https://www.researchgate.net/publication264933568_Software_Strategy_for_Reuse_Study_Final_Report/citations
9. Software Engineering Institute. (1990). *Establishing a basis for software reuse.* Retrieved from: https://insights.sei.cmu.edu/history-of-innovation/establishing-a-basis-for-software-reuse/#:~:text=Systematic%20software%20reuse%20is%20a%20strategy%20that,development%20environment%20brought%20about%20by%20budgetary%20restrictions
10. Schenkelberg, F. (2024). *Software reuse considerations.* Retrieved from: https://accendoreliability.com/software-reuse-considerations/?utm_source=chatgpt.com
11. Davenport, T. H., & Ronanki, R. (2018). *Artificial intelligence for the real world.* Deloitte.
12. INCOSE. (2024). *What is specialty engineering?* Retrieved from: https://www.incose.org/docs/default-source/TWG-Documents/005-specialty-engineering-and-se-pamphlet.pdf?sfvrsn=5cc882c6_4&utm_source=chatgpt.com
13. Dür, H., & Jones, E. (2012, September 30). *Building an Apollo 13 LiOH Canister Adapter.* Retrieved from https://www.nasa.gov/history/alsj/a13/a13_LIOH_Adapter.html?utm_source=chatgpt.com
14. Hansen, L. K., & Svejgaard Pold, S. (2023). *Stack bricolage and infrastructural impermanence in financial machine learning modelling.* Retrieved from https://research.cbs.dk/files/100410972/hansen_et_al_stack_bricolage_and_infrastructural_impermanence_acceptedversion.pdf
15. Olwell, D., Turner, R., Fairley, D., Jackson, S., & Squires, A. (n.d.). Systems engineering and specialty engineering. In *Guide to the systems engineering body of knowledge (SEBoK).* Retrieved from: https://sebokwiki.org/wiki/Systems_Engineering_and_Specialty_Engineering
16. Wang, Y. (2008). A hierarchical abstraction model for software engineering. In *Proceedings of the 6th ACM International Conference on Computing Frontiers* (pp. 1–8). ACM. https://doi.org/10.1145/1370164.1370174
17. Requisite Variety, Science Direct. (2024). Retrieved from: https://www.sciencedirect.com/topics/computer-science/requisite-variety
18. Hochstein, L. (2022, September 17). *Up and down the abstraction hierarchy. Surfing complexity.* Retrieved from https://surfingcomplexity.blog/2022/09/17/up-and-down-the-abstraction-hierarchy/
19. Blanchard, B. S., & Blyler, J. E. (2016). *System engineering management* (5th ed., p. 75). Wiley.
20. Kasser, J. (2014). *Managing complexity: The nine-system model. INCOSE Insight.* Retrieved from: https://www.researchgate.net/publication/273331086_Managing_complexity_The_Nine-System_Model
21. Patel, N. V. (2014). *The greatest space hack ever. Popular science.* Retrieved from: https://www.popsci.com/article/technology/greatest-space-hack-ever/#:~:text=To%20save%20themselves,%20the%20astronauts,tube%20socks,%20and%20duct%20tape

22. Wang, X., Zhang, Z., & Chun, D. (2025). Resource bricolage, digital transformation, and business model innovation: Based on the conditional process analysis of entrepreneurship systems. *Systems, 13*(5), 355. https://doi.org/10.3390/systems13050355
23. Deng, F., Ren, J., & Chen, F. (2018). *Abstract learning*. arXiv. Retrieved from: https://arxiv.org/abs/1809.03956v1
24. Kasser, J., & Zhao, Y. (2024). Managing complexity: The nine-system model. *INCOSE International Symposium*. Retrieved from: https://www.researchgate.net/publication/273331086
25. Jordan, M. I., & Mitchell, T. M. (2015). *Machine learning: Trends, perspectives, and prospects. Science, 349*(6245), 255–260. Retrieved from: https://www.cs.cmu.edu/~tom/pubs/Science-ML-2015.pdf
26. B. M., Ullman, T. D., Tenenbaum, J. B., & Gershman, S. J. (2016). Building machines that learn and think like people. *Center for Brains, Minds & Machines (CBMM) Memo No. 046*. arXiv https://doi.org/10.1017/S0140525X16001837
27. Holland, J. H. (1998). *Emergence: From chaos to order*. Oxford University Press. Retrieved from: https://www.scribd.com/document/47597042/Emergence-A-Critical-Analysis-of-Emergence-From-Chaos-to-Order-John-H-Holland
28. Simon, H. A. (1996). *The sciences of the artificial* (3rd ed.). MIT Press. Retrieved from: https://monoskop.org/images/9/9c/Simon_Herbert_A_The_Sciences_of_the_Artificial_3rd_ed.pdf
29. Alexander, C. (1977). *A pattern language: Towns, buildings, construction*. Oxford University Press. Retrieved from: https://www.academia.edu/39102604/
30. Beckett, R. C. (2016). *Entrepreneurial bricolage—Developing recipes to support innovation. International Journal of Innovation Management*. Retrieved from https://www.worldscientific.com/doi/abs/https://doi.org/10.1142/S1363919616400107.
31. Büscher, M., Gill, S., Mogensen, P., & Shapiro, D. (2010). Collective design: Bricolage and boundary objects in mobile ICT development. *Computer Supported Cooperative Work (CSCW), 19*(3–4), 245–281.
32. Moens, H., & De Turck, F. (2013). A scalable approach for structuring large-scale hierarchical cloud management systems. In *Conference on Network and Service Management* (pp. 1–8). Retrieved from https://www.semanticscholar.org/paper/A-scalable-approach-for-structuring-large-scale-Moens-Turck/ed0141a753b1a00742e4d2a7f965fe2b1217dfe5. https://doi.org/10.1109/CNSM.2013.6727803. 6.0 Path Forward: Developing Complex Projects through Abstraction and Hierarchy.
33. National Research Council. (2011). Recapturing a future for space exploration: Life and physical sciences research for a new era. The National Academies Press. Retrieved from https://nap.nationalacademies.org/download/13048.
34. Vaughan, D. (1996). The Challenger launch decision: Risky technology, culture, and deviance at NASA. University of Chicago Press. Retrieved from https://search.worldcat.org/search?q=bn:9780226346823.
35. Russell, S., & Norvig, P. (2020). Artificial intelligence: A modern approach (4th ed.). Pearson Education.
36. Deuze, M. (2006). Participation, remediation, bricolage: Considering principal components of a digital culture. The Information Society, 22, 63–75. https://doi.org/10.1080/01972240600567170. Retrieved from https://www.researchgate.net/publication/.
37. Jacobs, J. (1961). *The death and life of great American cities*. Random House. Retrieved from https://archive.org/details/deathlifeofgreat0000jaco_n0t5.
38. Holland, J. H. (1998). *Emergence: From chaos to order*. Oxford University Press. Retrieved from https://archive.org/details/emergencefromcha0000holl/mode/2up?q=emergence+.
39. Alexander, C. (1977). A Pattern Language: Towns, Buildings, Construction. Oxford University Press. Retrieved from:https://www.academia.edu/39102604/

40. Holland, J. H. (1998). *Emergence: From chaos to order.* Oxford University Press. Retrieved from https://www.researchgate.net/publication/375270833_Emergence_From_Chaos_to_Order
41. Gamma, E., Helm, R., Johnson, R., & Vlissides, J. (1993). Design patterns: Abstraction and reuse of object-oriented design. In *Lecture Notes in Computer Science* (Vol. 707, pp. 406–431). Springer. https://doi.org/10.1007/978-3-642-48354-7_15
42. DOOR3. (2023). *Integrated system design: Creating harmony in technology.* Retrieved from https://www.door3.com/blog/integrated-systems-design?utm_source=chatgpt.com

6 Path Forward: Developing Complex Projects Through Abstraction and Hierarchy

Establishing clear and flexible guidelines, heuristics, or rules to address development challenges as they arise is crucial for successfully managing a complex project. Abstraction and hierarchy are two practical approaches supporting project or system development analysis.

Abstraction allows us to simplify and generalize our understanding of a problem, while hierarchy helps us structure our knowledge and dependencies. The ability to abstract complex systems and concepts into more manageable representations is a foundational principle in programming and software design [1].

The hierarchical organization of projects enables a clearer understanding of the relationships and interdependencies between various system elements [2]. This understanding improves as heuristics are refined with every new project and development. At its core, developers craft the experience.

Abstraction is considered an important technique in systems engineering that helps manage complexity in large-scale projects. Focusing on core functions and isolating required elements makes creating scalable and efficient models easier. This approach allows engineers to move from detailed components to broader structures, emphasizing critical system behaviors, interfaces, and integration points, which leads to a more straightforward development process.

Hierarchy is a characteristic of complex system development due to its layered structure, which organizes components into manageable levels, identifies relationships and dependencies, and supports efficient planning and control. This structured organization makes it easier to understand system interactions and how different project components should fit into the larger system framework [2, 3].

F. A. Macias, *Project Development Simplified*, Synthesis Lectures on Engineering, Science, and Technology, https://doi.org/10.1007/978-3-032-01043-8_6

Establishing a clear hierarchical framework enables us to coordinate activities, identify interdependencies, and allocate resources more effectively. Research has shown that adopting a hierarchical approach can significantly improve the success rate of complex projects [2]. By breaking down the project into manageable sub-tasks and clearly defining the interfaces between them, project managers can better control the overall complexity and minimize the risk of cascading failures. Temporal relationships must be included in project developments, connecting all the management, design, and development strategies into a coherent framework [2].

Navigating complex project challenges is an important opportunity to develop and refine personal heuristics and rulesets continuously. These insights assist in making informed decisions, anticipating potential issues, and adapting strategies as the project evolves. Heuristics, or "rules of thumb," can be particularly useful when optimal solutions are challenging to determine or when time and resources are limited. By distilling past experiences and lessons into practical guidelines, project complexities can be navigated more effectively.

The unpredictability inherent in large-scale projects can lead to cost overruns and delays, often due to unforeseen problems that arise as the project progresses [4]. To mitigate these risks, establishing a comprehensive understanding of the project's complexity and implementing a resilient framework of heuristics and rulesets is crucial for effective decision-making and optimization [3].

Establishing a clear hierarchy, identifying critical dependencies, and maintaining a flexible approach to problem-solving are all key elements in developing successful complex projects. Developing complex projects requires a systematic approach that leverages the principles of abstraction and hierarchy. Abstraction allows us to simplify and generalize our understanding of the problem, while hierarchy helps us structure our knowledge and dependencies. By employing these strategies, project managers can navigate the challenges of complex endeavors more effectively and increase the likelihood of successful project outcomes [2–4].

Various research studies support the importance of abstraction and hierarchy in complex systems engineering. For instance, one study found that the combination of project manager leadership competencies and project complexity factors, such as structural, social, and emergent complexity, can significantly impact mega-projects' success [5]. Another study emphasized the need to regard projects as hierarchies, where management, design, and construction strategies must be aligned and expressed in clearly defined tasks [2].

Additionally, research has shown that determining the appropriate complexity level of a project and establishing the required levels of project rigor can be crucial for project success. The application of cognitive mapping methodologies has also been explored to understand the broader attributes of project management beyond the more execution-oriented aspects [6].

Insights into understanding how systems are to be developed, tested, or realized require abstractions to be applied as precisely as possible. Such an approach is part of the development of cognitive agents, which may start with structure and dependencies. Abstraction and hierarchical organization are essential principles in managing complex projects [2, 7]. There is a constant need to evolve and adapt the abstractions and hierarchical structures as projects progress and new challenges emerge.

Applying abstraction and hierarchical structures can significantly improve the management and delivery of complex projects. Abstraction enables project teams to simplify and generalize the problem, focusing on the core elements and relationships rather than getting bogged down in the details.

On the other hand, hierarchical organization helps project managers understand the interdependencies and relationships between different project components. Some of the key benefits of applying abstraction and hierarchy in complex project management include: [3, 8]

- Enhanced understanding of the overall project context and its components.
- Improved ability to coordinate and allocate resources effectively.
- Improved understanding and modeling of the problem domain.

Abstraction enables project teams to create simplified models of the problem, focusing on the essential elements and relationships. This approach helps manage the inherent complexity of large-scale projects and develop more effective strategies.

On the other hand, hierarchy provides a structured way of organizing and visualizing the project, making it easier to understand the context and the relationships between different components. By combining abstraction and hierarchy, project managers can more effectively navigate the challenges of complex projects, ensuring better alignment, coordination, and, ultimately, successful project outcomes.

Developing complex projects takes more than just technical knowledge. A structured approach is necessary to effectively utilize tools such as abstraction and hierarchy in managing complexity. Abstraction focuses attention on the essential parts of a problem and filters out unnecessary detail. Hierarchy helps by organizing parts of the system into a clear structure, showing how they connect and depend on one another. This makes it easier to understand what is going on and to manage the work. These approaches help teams stay focused, share information more effectively, and keep large projects moving forward. These methods are important to the practice of systems engineering and project management because they enable developers to understand the broader context, manage dependencies effectively, and work more efficiently.

Systems Engineering Management builds on that structure, not just by planning and oversight but also by intentionally shaping how a system functions across every level. Additionally, there is a need to combine technical expertise and leadership to ensure that a system is not only built correctly but also meets performance, reliability, cost, and usability objectives. As a result, the systems engineer must ensure that all aspects of the project, including technology, resources, and processes, work together and remain aligned.

Projects rarely run smoothly. Plans change, requirements evolve, and miscommunication happens. That is where structure becomes more than just a theory. Abstraction and hierarchy enable engineers to stay focused, prioritize tasks, and adapt effectively. The best systems engineers know how to create order out of uncertainty. They utilize processes such as requirements management, system integration, risk assessment, and testing not only to stay on track but also to improve along the way continually.

The real challenge is managing complex systems in a way that goes beyond just building them. The goal is to design an experience that works for the Stakeholders who use it and the developers who build it.

References

1. Edwards, J., & Petříček, T. (2022). *Interaction vs. abstraction: Managed copy and paste.* https://www.subtext-lang.org/PAINT22.pdf
2. Walker, A. (1985). *Project management in construction* (p. 41). BSP Professional Books. https://www.google.com/books/edition/Project_Management_in_Construction/4Q51BgAAQBAJ?hl=en&gbpv=1
3. Andrews, T. D. (2015, October). *Using project complexity determinations to establish required levels of project rigor.* Paper presented at the PMI® Global Congress 2015—North America, Orlando, FL. Project Management Institute. https://www.pmi.org/learning/library/project-complexity-determine-rigor-9874
4. Madhavan, G., Romig, A. D., Meserve, R. A., & Winter, D. C. (2023). Delivering effectively on large engineering projects. *PNAS Nexus, 2*(9). Oxford University Press. https://doi.org/10.1093/pnasnexus/pgad281
5. Damayanti, R. W., Hartono, B., & Wijaya, A. R. (2021). Complexity, leadership, and megaproject performance: A configuration analysis. *Journal of Industrial Engineering and Management, 14*(3), 570. OmniaScience. https://doi.org/10.3926/jiem.3476
6. Edkins, A., Kurul, E., Maytorena-Sanchez, E., & Rintala, K. (2007). The application of cognitive mapping methodologies in project management research. *International Journal of Project Management, 25*(8), 762–772. Elsevier BV. https://www.sciencedirect.com/science/article/pii/S0263786307000713?via=ihub
7. Daniels, C. B., & LaMarsh, W. J. (2007). Complexity as a cause of failure in information technology project management. In *Proceedings of the 2007 IEEE international conference on system of systems engineering* (pp. 1–6). IEEE. https://doi.org/10.1109/SYSOSE.2007.4304225.
8. Morrison, J. M., & Brown, C. (2004). Project management effectiveness as a construct: A conceptual study. *South African Journal of Business Management, 35*(4), 73–83. AOSIS. https://doi.org/10.4102/sajbm.v35i4.670

MIX
Papier aus verantwortungsvollen Quellen
Paper from responsible sources
FSC® C105338

If you have any concerns about our products,
you can contact us on
ProductSafety@springernature.com

In case Publisher is established outside the EU,
the EU authorized representative is:
Springer Nature Customer Service Center GmbH
Europaplatz 3, 69115 Heidelberg, Germany

Printed by Libri Plureos GmbH
in Hamburg, Germany